HEINEMANN MODULAR MATHEMATICS
for
EDEXCEL AS AND A-LEVEL
Mechanics 3

John Hebborn   Jean Littlewood

Examination style paper M3

THE
HARRODIAN SCHOOL

SEP 200

Endorsed by **edexcel**

Heinemann is an imprint of Pearson Education Limited,
a company incorporated in England and Wales, having
its registered office at Edinburgh Gate, Harlow, Essex, CM20 2JE.
Registered company number: 872828

Heinemann is a registered trademark of Pearson Education Limited

First published 2000

08 9

ISBN 978 0 435510 76 3

Cover design by Gecko Limited

Original design by Geoffrey Wadsley: additional design work by Jim Turner

Typeset and illustrated by Tech-Set Ltd, Gateshead, Tyne and Wear

Printed in China by CTPS

Acknowledgements:

The publisher's and authors' thanks are due to Edexcel for permission to
reproduce questions from past examination papers. These are marked with an [E].
   The answers have been provided by the authors and are not the responsibility
of the examining board.

About this book

This book is designed to provide you with the best preparation possible for your Edexcel M3 exam. The series authors are senior examiners and exam moderators themselves and have a good understanding of Edexcel's requirements.

Finding your way around

To help you find your way around when you are studying and revising use the:

- **edge marks** (shown on the front page) – these help you to get to the right chapter quickly;
- **contents list** – this lists the headings that identify key syllabus ideas covered in the book so you can turn straight to them;
- **index** – if you need to find a topic the **bold** number shows where to find the main entry on a topic.

Remembering key ideas

We have provided clear explanations of the key ideas and techniques you need throughout the book. Key ideas you need to remember are listed in a **summary of key points** at the end of each chapter and marked like this in the chapters:

$$e = \frac{v}{u} \text{ or } v = eu$$

Exercises and exam questions

In this book questions are carefully graded so they increase in difficulty and gradually bring you up to exam standard.

- **past exam questions** are marked with an E;
- **review exercises** on pages 97 and 209 help you practise answering questions from several areas of mathematics at once, as in the real exam;
- **exam style practice paper** – this is designed to help you prepare for the exam itself;
- **answers** are included at the end of the book – use them to check your work.

Contents

Further kinematics

Chapter 1 of book M2 dealt with the kinematics of a particle when the acceleration is constant. Remember that the constant acceleration formulae derived in section 3.1 of Book M1 must only be used when the acceleration is constant. However, in practice, the acceleration of a particle is often not constant. This chapter revises and extends the material covered in section 1.2 of Book M2 and considers the kinematics of a particle moving in one dimension, when the acceleration is not constant.

Two cases will be considered:

(1) when the acceleration depends on time,
(2) when the acceleration depends on the displacement.

In both cases calculus methods must be used.

1.1 Acceleration depending on time (*t*)

Consider a particle P moving in a straight line. Suppose that at time t seconds the displacement of P from a fixed point O in the line is x metres, its velocity is $v\,\mathrm{m\,s}^{-1}$ and its acceleration is $a\,\mathrm{m\,s}^{-2}$.

You will recall from chapter 1 of book M2 that: $a = \dfrac{\mathrm{d}v}{\mathrm{d}t}$

and: $v = \dfrac{\mathrm{d}x}{\mathrm{d}t}$

so: $a = \dfrac{\mathrm{d}^2x}{\mathrm{d}t^2}$

When the acceleration is a function of time:
$$a = \mathrm{f}(t)$$

So: $v = \displaystyle\int \mathrm{f}(t)\mathrm{d}t + c$

where c is an arbitrary constant of integration.

Similarly, when $v = g(t)$:

then: $$x = \int g(t)\mathrm{d}t + k$$

where k is an arbitrary constant of integration.

Example 1

A particle P is moving along the line Ox with a constant acceleration equal to $A\,\mathrm{m\,s}^{-2}$ ($A > 0$). When $t = 0$ the particle is at O, so that $x = 0$, and it has a speed of $u\,\mathrm{m\,s}^{-1}$ in the direction Ox.

(a) Obtain, by using calculus, v and x as functions of t.
(b) Sketch the graphs of v against t and x against t.

(a) As: $$a = \frac{\mathrm{d}v}{\mathrm{d}t} = A$$

So: $$v = \int A\mathrm{d}t + c$$

$$= At + c, \text{ since } A \text{ is constant.}$$

The value of the constant c may be found by using $v = u$ when $t = 0$.

Then: $$u = 0 + c$$

$\Rightarrow$ $$c = u$$

and: $$v = u + At$$

As: $$v = \frac{\mathrm{d}x}{\mathrm{d}t} = u + At$$

so: $$x = \int (u + At)\mathrm{d}t + k$$

$$= ut + \tfrac{1}{2}At^2 + k, \text{ since both } A \text{ and } u \text{ are constant.}$$

As $x = 0$ when $t = 0$ the constant k may be found.

Since: $$0 = 0 + 0 + k$$

then: $$k = 0$$

and so: $$x = ut + \tfrac{1}{2}At^2$$

(b) The sketch of v against t is a straight line with gradient A which cuts the v-axis at $v = u$.

To sketch x against t note:

(1) The gradient of the curve is $(u + At)$ and this is always positive for $t > 0$, as both u and A are positive. The gradient is zero when $t = -\dfrac{u}{A}$.

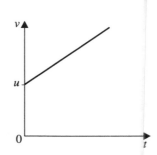

(2) $x = 0$ when $ut + \frac{1}{2}At^2 = 0$

That is, when:
$$t(u + \tfrac{1}{2}At) = 0$$

So:
$$t = 0 \text{ or } t = -\frac{2u}{A}$$

So for $t \geqslant 0$ the only solution is $t = 0$.

(3) When $t = 0$, $x = 0$ so the curve goes through (0,0). The sketch of x against t may now be drawn.

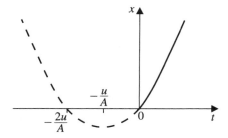

Example 2

A particle P is moving along the line Ox with an acceleration $\frac{2}{3}t\,\mathrm{m\,s^{-2}}$ at time t seconds. When $t = 3$ the speed of P is $4\,\mathrm{m\,s^{-1}}$ in the direction Ox and its displacement from O is $10\,\mathrm{m}$. Find the displacement of P from O when $t = 6$.

As:
$$a = \frac{\mathrm{d}v}{\mathrm{d}t} = \frac{2}{3}t$$

so:
$$v = \int \frac{2}{3}t\,\mathrm{d}t + c$$

$$= \frac{2}{3}\left(\frac{t^2}{2}\right) + c = \frac{1}{3}t^2 + c$$

where c is an arbitrary constant.

Since $v = 4$ when $t = 3$
$$4 = \frac{1}{3}(3)^2 + c$$

So:
$$c = 1$$

and:
$$v = \frac{1}{3}t^2 + 1$$

From:
$$v = \frac{\mathrm{d}x}{\mathrm{d}t} = \frac{1}{3}t^2 + 1$$

integrating with respect to t gives:

$$x = \int \left(\frac{1}{3}t^2 + 1\right) dt + k$$

$$= \frac{1}{3}\left(\frac{t^3}{3}\right) + t + k$$

$$= \frac{1}{9}t^3 + t + k$$

where k is an arbitrary constant.

Since $x = 10$ when $t = 3$

$$10 = \frac{1}{9}(3)^3 + 3 + k$$

So: $\qquad k = 4$

and: $\qquad x = \frac{1}{9}t^3 + t + 4$

When $t = 6$: $\qquad x = \frac{1}{9}(6)^3 + 6 + 4 = 34$

The displacement of P from O when $t = 6$ is $34\,\text{m}$.

Example 3

A particle P is moving along the line Ox with an acceleration $(6t - 18)\,\text{m s}^{-2}$ at time t seconds. When $t = 0$ the displacement of P from O is zero and its speed is $24\,\text{m s}^{-1}$ in the direction Ox.

(a) Find the values of t for which $v = 0$ and the values of x for these two values of t.

(b) Find the distance travelled in the first 3 seconds.

On first sight this question may appear to be exactly the same as example 2. There are however some important differences. In example 2 the acceleration was only zero when $t = 0$ and was positive for all $t > 0$. The speed $v\,\text{m s}^{-1}$ never vanished – it was in fact always positive and so x increased with t.

In the solution the differences between this example and example 2 will be highlighted.

(a) As: $\qquad a = \dfrac{\mathrm{d}v}{\mathrm{d}t} = 6t - 18$

integrating with respect to t gives:

$$v = \int (6t - 18)\mathrm{d}t + c$$

$$= 6\left(\frac{t^2}{2}\right) - 18t + c$$

$$= 3t^2 - 18t + c$$

where c is an arbitrary constant.

Since $v = 24$ when $t = 0$

$$24 = 0 - 0 + c$$

So:
$$c = 24$$

and:
$$v = 3t^2 - 18t + 24 \qquad (1)$$

The values of t for which $v = 0$ are obtained from

$$3t^2 - 18t + 24 = 0$$

or:
$$t^2 - 6t + 8 = 0$$

Factorising:
$$(t - 4)(t - 2) = 0$$

so:
$$t = 2 \quad \text{or} \quad t = 4$$

It is useful to sketch v against t for this situation as this gives us a clear picture of how v varies with time. The following facts will enable us to produce this sketch.

(1) $v = 0$ when $t = 2$ and $t = 4$

(2) when $t = 0$, $v = 24$

(3) v has a stationary value when $\dfrac{dv}{dt}$, that is a, is zero (see section 5.7 of Book P1). So this occurs when $6t - 18 = 0$, that is when $t = 3$.

The sketch is:

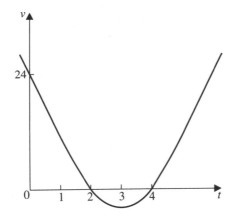

To find the distances travelled by P when $t = 2$ and $t = 4$ use:

$$v = \frac{dx}{dt} = 3t^2 - 18t + 24 \text{ from equation (1).}$$

Integrating with respect to t:

$$x = 3\left(\frac{t^3}{3}\right) - 18\left(\frac{t^2}{2}\right) + 24t + k$$

$$= t^3 - 9t^2 + 24t + k$$

where k is an arbitrary constant.

Since $x = 0$ when $t = 0$

$$0 = 0 - 0 + 0 + k$$

So: $\qquad\qquad\qquad\quad k = 0$

and: $\qquad\qquad\qquad x = t^3 - 9t^2 + 24t \qquad\qquad (2)$

When $t = 2$ the value of x is given by:

$$x = (2)^3 - 9(2)^2 + 24(2)$$

So $x = 20$ m and P is at a point A.

When $t = 4$ the value of x is given by:

$$x = (4)^3 - 9(4)^2 + 24(4)$$

So $x = 16$ m and P is at a point B.

These solutions are illustrated below on the line Ox together with the initial situation.

Notice that the displacement of P from O when $t = 4$ is smaller than the displacement of P from O when $t = 2$. This is because v is negative for $2 < t < 4$, as shown in the sketch of v against t, and since $v = \dfrac{dx}{dt}$, x must be decreasing for $2 < t < 4$.

(b) The displacement of P from O when $t = 3$ is obtained by substituting $t = 3$ into equation (2).

Hence: $\qquad\qquad x = (3)^3 - 9(3)^2 + 24(3)$

$$= 18$$

So $x = 18$ m and P is at a point C between A and B.

In the first two seconds v is positive and P travels from O to A a distance of 20 m. In the third second v is negative so the direction of motion is reversed and P moves from A to C, a distance of $(20 - 18)$ m $= 2$ m.

The total distance travelled in the first 3 seconds is therefore

$$OA + AC = 20\,\text{m} + 2\,\text{m}$$

$$= 22\,\text{m}$$

Example 4

A particle P starts from rest at O at time $t = 0$ and moves with variable acceleration a m s^{-2} where

$$a = \frac{t}{5}, \qquad 0 \leqslant t < 5$$

$$a = \frac{t}{5} + \frac{10}{t^2}, \qquad t \geqslant 5$$

where t is measured in seconds. Show that the speed of P is $2\frac{1}{2}$ m s^{-1} when $t = 5$ and 11 m s^{-1} when $t = 10$. Show that the displacement of P from O when $t = 10$ is $(43\frac{1}{3} - 10\ln 2)$ metres.

As a has a different form in the two regions these regions must be considered separately. The fact that v and x must be continuous at $t = 5$, the boundary between the two regions, must then be used.

For $0 \leqslant t < 5$ (region I)

$$a = \frac{dv}{dt} = \frac{t}{5}$$

So, integrating with respect to t:

$$v = \int \frac{t}{5}\, dt = \frac{1}{5}\left(\frac{t^2}{2}\right) + c = \frac{t^2}{10} + c$$

where c is an arbitrary constant. The value of c can be found since P starts from rest at $t = 0$.

Hence: $\qquad v = 0$ at $t = 0 \Rightarrow 0 = 0 + c$

So: $\qquad c = 0$

and: $\qquad v = \frac{t^2}{10}, 0 \leqslant t < 5$ $\qquad\qquad$ (1)

Hence when $t = 5$: $\qquad v = \frac{5^2}{10} = 2\frac{1}{2}$

The speed of P is $2\frac{1}{2}$ m s^{-1} when $t = 5$.

For $t \geqslant 5$ (region II)

$$a = \frac{dv}{dt} = \left(\frac{t}{5} + \frac{10}{t^2}\right)$$

So, integrating with respect to t:

$$v = \int \left(\frac{t}{5} + \frac{10}{t^2}\right) dt = \frac{t^2}{10} + \frac{10t^{-1}}{(-1)} + k$$

$$= \frac{t^2}{10} - \frac{10}{t} + k$$

where k is an arbitrary constant.

To find k use the result obtained above, that is $v = 2\frac{1}{2}$ when $t = 5$.

So: $$2\frac{1}{2} = \frac{25}{10} - \frac{10}{5} + k$$

and: $$k = 2$$

Hence: $$v = \frac{t^2}{10} - \frac{10}{t} + 2, \quad t \geqslant 5 \qquad\qquad (2)$$

When $t = 10$ we have:
$$v = \frac{100}{10} - \frac{10}{10} + 2 = 11$$

The speed of P is $11\,\mathrm{m\,s}^{-1}$ when $t = 10$.

To find the displacement of P you must return to equation (1). In region I for $0 \leqslant t < 5$:
$$v = \frac{\mathrm{d}x}{\mathrm{d}t} = \frac{t^2}{10}$$

So, integrating with respect to t gives:
$$x = \frac{1}{10}\left(\frac{t^3}{3}\right) + A$$

where A is an arbitrary constant. Taking $x\,\mathrm{m}$ to be the displacement from O and using the condition $x = 0$ at $t = 0$, since P starts from O, gives:
$$0 = 0 + A$$

So: $$A = 0$$

and: $$x = \frac{t^3}{30}, \quad 0 \leqslant t < 5$$

When $t = 5$ you find that $x = \dfrac{5^3}{30} = 4\frac{1}{6}\mathrm{m}$.

In region II, using equation (2), for $t \geqslant 5$
$$v = \frac{\mathrm{d}x}{\mathrm{d}t} = \frac{t^2}{10} - \frac{10}{t} + 2$$

So, integrating with respect to t gives:
$$x = \frac{1}{10}\left(\frac{t^3}{3}\right) - 10\ln t + 2t + B$$
$$= \frac{t^3}{30} - 10\ln t + 2t + B$$

where B is an arbitrary constant.

From above, $x = 4\frac{1}{6}$ when $t = 5$ and substituting these values gives:
$$4\frac{1}{6} = \tfrac{1}{30}(5)^3 - 10\ln 5 + 10 + B$$

$\Rightarrow \qquad\qquad B = 10\ln 5 - 10$

Hence: $x = \dfrac{t^3}{30} - 10\ln t + 2t + 10\ln 5 - 10, \quad t \geq 5$

When $t = 10$:

$$x = \frac{(10)^3}{30} - 10\ln 10 + 20 + 10\ln 5 - 10$$
$$= 43\tfrac{1}{3} - 10\ln 2$$

The displacement of P from O when $t = 10$ is $(43\tfrac{1}{3} - 10\ln 2)$ metres.

Example 5

A particle P moves in a straight line with acceleration $(\sin \omega t)\,\mathrm{m\,s}^{-2}$ at time t seconds where ω is constant. Initially the particle is at rest and its displacement from a fixed point O on the line is 1 metre.

(a) Find the velocity of P at any time t.
(b) Find the displacement of P from O at any time t.
(c) Find the value of t when P next comes to instantaneous rest.

(a) Since: $\qquad a = \dfrac{dv}{dt} = \sin \omega t$

so: $\qquad v = \displaystyle\int \sin \omega t\, dt + c$

$$= -\frac{1}{\omega}\cos \omega t + c$$

where c is an arbitrary constant.

Since $v = 0$ when $t = 0$:

$$0 = -\frac{1}{\omega}\cos 0 + c$$

So: $\qquad c = \dfrac{1}{\omega}$, as $\cos 0 = 1$,

and: $\qquad v = \dfrac{1}{\omega}(1 - \cos \omega t)$

(b) As: $\qquad v = \dfrac{dx}{dt} = \dfrac{1}{\omega}(1 - \cos \omega t)$

so: $\qquad x = \dfrac{1}{\omega}\displaystyle\int (1 - \cos \omega t)dt + k$

$$= \frac{1}{\omega}\left(t - \frac{1}{\omega}\sin \omega t\right) + k$$

where k is an arbitrary constant.

Since $x = 1$ when $t = 0$:

$$1 = 0 - 0 + k$$

So: $\qquad k = 1$

and: $\qquad x = \dfrac{t}{\omega} - \dfrac{1}{\omega^2}\sin \omega t + 1$

(c) *P* comes to rest when $v = 0$, that is:

$$\frac{1}{\omega}(1 - \cos \omega t) = 0$$

$\Rightarrow$ $$\cos \omega t = 1$$

$$\omega t = 2m\pi, \text{ where } m \text{ is an integer}$$

So: $$t = 0, \frac{2\pi}{\omega}, \frac{4\pi}{\omega}, \ldots$$

P next comes to rest when $t = \dfrac{2\pi}{\omega}$ s.

Example 6

A particle *P* moves in a straight line with acceleration $(\cos \pi t)\,\text{m s}^{-2}$ at time *t* seconds. The particle starts from rest at $t = 0$. Obtain the distance travelled by *P* in the time interval from $t = 1$ to $t = 2$.

As: $$a = \frac{dv}{dt} = \cos \pi t$$

so: $$v = \int \cos \pi t \, dt + c$$

$$= \frac{1}{\pi} \sin \pi t + c$$

Since $v = 0$ when $t = 0$:

$$0 = \frac{1}{\pi} \sin 0 + c$$

So: $$c = 0, \text{ as } \sin 0 = 0,$$

and: $$v = \frac{1}{\pi} \sin \pi t$$

Before proceeding it is useful to sketch *v* against *t*. The graph of $y = \sin x$ is shown in Book P1, section 2.5 and using this gives:

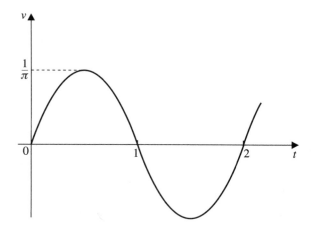

For $1 \leqslant t \leqslant 2$ the velocity is never positive which means that P is moving in the same direction throughout this interval – in the negative x direction.

To obtain information about displacements and distances integrate again with respect to t.

As:
$$v = \frac{dx}{dt} = \frac{1}{\pi} \sin \pi t$$

so:
$$x = \frac{1}{\pi} \int (\sin \pi t) dt + k$$

$$= -\frac{1}{\pi^2} \cos \pi t + k$$

where k is an arbitrary constant.

As no information is given about the displacement at some fixed time the value of k cannot be found. However, as you will see below, it is not necessary to do so.

When $t = 1$ the displacement:
$$x_1 = -\frac{1}{\pi^2} \cos \pi + k$$

$$= +\frac{1}{\pi^2} + k, \text{ as } \cos \pi = -1$$

When $t = 2$ the displacement:
$$x_2 = -\frac{1}{\pi^2} \cos 2\pi + k$$

$$= -\frac{1}{\pi^2} + k, \text{ as } \cos 2\pi = 1$$

As the motion in the given interval is in a constant direction, the distance travelled by P is given by

$$|x_2 - x_1| = \left| \left(-\frac{1}{\pi^2} + k \right) - \left(\frac{1}{\pi^2} + k \right) \right|$$

$$= \frac{2}{\pi^2}$$

Notice that k does not appear in the answer as it cancels out.

The distance travelled by P in the time interval from $t = 1$ to $t = 2$ is $\frac{2}{\pi^2}$ metres.

Example 7

A particle P moves in a straight line with acceleration $(12t) \, \text{m s}^{-2}$ at time t seconds. Its displacement at this time is x metres. Given that $x = 5$ when $t = 1$ and $x = 59$ when $t = 3$, find an expression for x as a function of t and also the value of x when $t = 4$.

As:
$$a = \frac{dv}{dt} = 12t$$

so:
$$v = \int 12t\,dt + A$$
$$= 12\left(\frac{t^2}{2}\right) + A = 6t^2 + A$$

where A is an arbitrary constant.

No information is given about the value of v at any time so you cannot find A yet.

Integrating again with respect to t:
$$x = \int (6t^2 + A)dt + B$$
$$= 6\left(\frac{t^3}{3}\right) + At + B$$
$$= 2t^3 + At + B$$

where B is another arbitrary constant.

Since $x = 5$ when $t = 1$:
$$5 = 2 + A + B$$

So:
$$A + B = 3 \qquad (1)$$

Since $x = 59$ when $t = 3$:
$$59 = 2(3)^3 + 3A + B$$
$$= 54 + 3A + B$$

So:
$$3A + B = 5 \qquad (2)$$

Solve equations (1) and (2) to obtain A and B.

Subtracting equation (1) from equation (2) gives:
$$2A = 2$$

So:
$$A = 1$$

Substituting into equation (1):
$$1 + B = 3$$
$$\Rightarrow B = 2$$

Hence:
$$x = 2t^3 + t + 2$$

When $t = 4$:
$$x = 2(4)^3 + 4 + 2$$
$$= 128 + 4 + 2$$
$$x = 134$$

When $t = 4$ s, $x = 134$ m.

Exercise 1A

1 A particle P moves in a straight line with an acceleration $4t \, \mathrm{m \, s^{-2}}$ at time t seconds. When $t = 0$ the velocity of P is $3 \, \mathrm{m \, s^{-1}}$. Find the velocity of P when $t = 3$.

2 A particle P moves in a straight line with an acceleration $6t \, \mathrm{m \, s^{-2}}$ at time t seconds. The particle starts from rest. Find the velocity of P when $t = 4$.

3 A particle P moves in a straight line with an acceleration $2t \, \mathrm{m \, s^{-2}}$ at time t seconds. It starts from rest at O. Find the velocity of P and the distance of P from O when $t = 3$.

4 A particle P moves in a straight line with acceleration $\left(\dfrac{3t^2}{2}\right) \mathrm{m \, s^{-2}}$ at time t seconds. The particle passes through the point O when $t = 2$ with velocity $1 \, \mathrm{m \, s^{-1}}$. Find the velocity of P when $t = 4$ and the distance of P from O at this time.

5 A particle P starts from rest at O at time $t = 0$ and moves with variable acceleration $a \, \mathrm{m \, s^{-2}}$ where

$$a = 2t, \quad 0 \leqslant t < 3$$
$$a = -2, \quad \quad t \geqslant 3$$

where t is measured in seconds. Find the value of t when the particle comes to rest and the displacement of P from O at this time.

6 A particle P moves in a straight line with an acceleration $(3t - 1) \, \mathrm{m \, s^{-2}}$ at time t seconds. The particle has a velocity of $3 \, \mathrm{m \, s^{-1}}$ when $t = 2$. Find the velocity of P when $t = 4$.

7 A particle P moves in a straight line with an acceleration $(3t - 6) \, \mathrm{m \, s^{-2}}$ at time t seconds. Initially P is at a distance of $2 \, \mathrm{m}$ from O and moving with a velocity of $4\frac{1}{2} \mathrm{m \, s^{-1}}$.
(a) Find the values of t when the velocity is zero.
(b) Find the displacement of the particle from O when $t = 4$.

8 A particle P moves in a straight line with acceleration $(\sin 2t) \, \mathrm{m \, s^{-2}}$ at time t seconds. The particle is initially at rest at O. Find the velocity of P and its displacement from O when $t = \dfrac{\pi}{2}$.

9 A particle moves in a straight line and its acceleration after
   t seconds is $6t \, \mathrm{m \, s^{-2}}$. Its displacement at time t is x metres.
   Given that $x = 5$ when $t = 1$ and $x = 14$ when $t = 2$, find an
   expression for x as a function of t and the value of x when $t = 3$.

10 A particle P moves along Ox with an acceleration $(e^{-t}) \, \mathrm{m \, s^{-2}}$
   at time t seconds. Given that at $t = 0$ the speed of P is $4 \, \mathrm{m \, s^{-1}}$
   in the direction Ox, and its displacement from O is $5 \, \mathrm{m}$, find
   the displacement of P from O as a function of t.

11 A particle P moves along Ox with an acceleration
   $\dfrac{-1}{(4 + t)^{1/2}} \, \mathrm{m \, s^{-2}}$ at time t seconds. Given that at $t = 0$ the
   speed of P is $2 \, \mathrm{m \, s^{-1}}$ in the direction Ox, find the value of t
   when the speed of P is zero.

12 A particle P moves along Ox with an acceleration $\dfrac{3}{(1 + t)^2} \, \mathrm{m \, s^{-2}}$
   at time t seconds. The particle starts from rest at O at time

   $t = 0$. Find the displacement of P from O when $t = 6$.

13 A particle P moves in a straight line with acceleration, at time
   t seconds, proportional to $(t + t_0)^{-3}$, where t_0 is positive and
   constant. The initial speed of P at $t = 0$ is $u \, \mathrm{m \, s^{-1}}$.
   Show that the speed of P approaches a limiting value as $t \to \infty$.
   Given that this limiting value is $2u \, \mathrm{m \, s^{-1}}$, show that at time t
   seconds the particle will have travelled a distance
   $$\frac{ut(2t + t_0)}{(t + t_0)} \text{ metres.}$$

1.2 Acceleration depending
on displacement (x)

Consider a particle P moving in a straight line. Suppose that at time
t seconds the displacement of P from a fixed point O in the line is
x metres, its velocity is $v \, \mathrm{m \, s^{-1}}$ and its acceleration is $a \, \mathrm{m \, s^{-2}}$.

In section 1.1 the equation $a = \dfrac{\mathrm{d}v}{\mathrm{d}t}$ was used. When the acceleration
is a function of the displacement:
$$a = \mathrm{F}(x)$$

so:
$$\frac{\mathrm{d}v}{\mathrm{d}t} = \mathrm{F}(x)$$

This equation contains 3 variables: x, t and v. In order to be able to solve this equation an alternative form for the acceleration must be obtained.

Write: $\quad a = \dfrac{\mathrm{d}v}{\mathrm{d}t} = \dfrac{\mathrm{d}v}{\mathrm{d}x}\dfrac{\mathrm{d}x}{\mathrm{d}t}$ by the chain rule (Book P3, chapter 2)

Since $\dfrac{\mathrm{d}x}{\mathrm{d}t} = v$ this gives:

∎
$$a = v\,\frac{\mathrm{d}v}{\mathrm{d}x}$$

Since $\dfrac{\mathrm{d}}{\mathrm{d}x}(v^2) = 2v\,\dfrac{\mathrm{d}v}{\mathrm{d}x}$ the acceleration may also be written:

∎
$$a = \frac{\mathrm{d}}{\mathrm{d}x}\left(\tfrac{1}{2}v^2\right)$$

Example 8

A particle P moves along the positive x-axis and its acceleration is $3x\,\mathrm{m\,s}^{-2}$, when its displacement from the origin O is x metres. Given that when $x = 0$ the velocity of P is $8\,\mathrm{m\,s}^{-1}$ in the direction Ox, obtain an expression for v in terms of x.

As: $\qquad\qquad\qquad\qquad a = \dfrac{\mathrm{d}}{\mathrm{d}x}\left(\tfrac{1}{2}v^2\right)$

so: $\qquad\qquad\qquad\qquad \dfrac{\mathrm{d}}{\mathrm{d}x}\left(\tfrac{1}{2}v^2\right) = 3x$

Integrating with respect to x gives:

$$\tfrac{1}{2}v^2 = \int 3x\,\mathrm{d}x + c$$
$$= 3\left(\frac{x^2}{2}\right) + c = 1\tfrac{1}{2}x^2 + c$$

where c is an arbitrary constant.

Since $v = 8$ when $x = 0$:

$$\tfrac{1}{2}(8)^2 = 0 + c$$

So: $\qquad\qquad\qquad\qquad c = 32$

and: $\qquad\qquad\qquad\qquad \tfrac{1}{2}v^2 = 1\tfrac{1}{2}x^2 + 32$

or: $\qquad\qquad\qquad\qquad v^2 = 3x^2 + 64$

Taking square roots:

$$v = \pm\sqrt{(3x^2 + 64)}$$

The acceleration given in the question is positive for $x > 0$ so that P will always be moving in the direction Ox and v will always be positive and increasing.

So: $\qquad\qquad\qquad\qquad v = \sqrt{(3x^2 + 64)}$

Example 9

A particle P moves along a straight line and its acceleration when it has a displacement of x metres from O, a fixed point in the line, is of magnitude $8x\,\mathrm{m\,s^{-2}}$ and directed towards O. The velocity of the particle is $v\,\mathrm{m\,s^{-1}}$ and initially $v = 0$ when $x = 2$.

(a) Obtain an expression for v^2 as a function of x.
(b) Find the values of x for which $v = 0$.
(c) Find the maximum speed of P.

(a) You are given $a = \dfrac{\mathrm{d}}{\mathrm{d}x}\left(\dfrac{1}{2}v^2\right) = -8x$

The minus sign shows that the acceleration is directed *towards* O.

So integrating with respect to x gives:

$$\frac{1}{2}v^2 = -\int 8x\,\mathrm{d}x + c$$

$$= -8\left(\frac{x^2}{2}\right) + c = -4x^2 + c$$

where c is an arbitrary constant.

Since $v = 0$ when $x = 2$:

$$0 = -4(2)^2 + c$$

So: $\qquad\qquad\qquad c = 16$

and: $\qquad\qquad\quad \frac{1}{2}v^2 = -4x^2 + 16$

or: $\qquad\qquad\quad v^2 = -8x^2 + 32 = 32 - 8x^2 \qquad\qquad (1)$

(b) Substituting $v = 0$ in equation (1) gives:

$$32 - 8x^2 = 0$$

or: $\qquad\qquad\qquad 4 - x^2 = 0$

so: $\qquad\qquad\qquad x^2 = 4$

$\Rightarrow \qquad\qquad\qquad x = \pm 2$

(c) The stationary values of v occur when:

$$a = \frac{\mathrm{d}v}{\mathrm{d}t} = v\frac{\mathrm{d}v}{\mathrm{d}x} = \frac{\mathrm{d}}{\mathrm{d}x}\left(\frac{1}{2}v^2\right) = 0$$

That is when: $\qquad 8x = 0$

$\Rightarrow \qquad\qquad\qquad x = 0$

When $x = 0$: $\qquad\qquad v^2 = 32$

$\Rightarrow \qquad\qquad\qquad v = \pm\sqrt{32} = \pm 4\sqrt{2}$

The maximum value of $|v|$ is $4\sqrt{2}$ and so the maximum speed of P is $4\sqrt{2}\,\mathrm{m\,s^{-1}}$.

Example 10

A particle P moves along the line Ox. When its displacement from O is x metres, its acceleration is of magnitude $\left(\dfrac{k^2}{2x^2}\right)$ m s^{-2} and directed towards O. Given that when $x = \frac{1}{4}a$ the velocity of P is $\left(\dfrac{3k^2}{a}\right)^{\frac{1}{2}}$ m s^{-1} in the direction Ox, find the value of x when P comes to rest.

In this case:

$$a = \frac{\mathrm{d}}{\mathrm{d}x}\left(\frac{1}{2}v^2\right) = -\frac{k^2}{2x^2}$$

as the acceleration is towards O.

So integrating with respect to x:

$$\frac{1}{2}v^2 = -\int \frac{k^2}{2x^2}\,\mathrm{d}x + c$$

$$= -\frac{k^2}{2}\left(\frac{x^{-1}}{-1}\right) + c$$

$$= \frac{k^2}{2x} + c$$

where c is an arbitrary constant.

As $v = \left(\dfrac{3k^2}{a}\right)^{\frac{1}{2}}$ when $x = \frac{1}{4}a$,

$$\frac{1}{2}\left(\frac{3k^2}{a}\right) = \frac{k^2}{2\left(\frac{a}{4}\right)} + c$$

So:

$$c = \frac{3k^2}{2a} - \frac{4k^2}{2a} = -\frac{k^2}{2a}$$

and:

$$\frac{1}{2}v^2 = \frac{k^2}{2x} - \frac{k^2}{2a}$$

or:

$$v^2 = k^2\left(\frac{1}{x} - \frac{1}{a}\right)$$

P comes to rest when $v = 0$, that is when:

$$\frac{1}{x} - \frac{1}{a} = 0$$

So:

$$x = a$$

When P comes to rest, $x = a$.

Example 11

A particle P moves along the line Ox. When its displacement from O is x metres, its acceleration is of magnitude $\left(\dfrac{8}{x^2}\right)$ m s^{-2} and directed towards O. When $x = 1$ the velocity of P is 5 m s^{-1} in the direction Ox. Show that the velocity of P tends to 3 as x tends to infinity.

In this case:
$$a = \frac{\mathrm{d}}{\mathrm{d}x}\left(\frac{1}{2}v^2\right) = -\frac{8}{x^2}$$

Integrating with respect to x:

$$\frac{1}{2}v^2 = \int -\frac{8}{x^2}\,\mathrm{d}x + c = -8\left(\frac{x^{-1}}{-1}\right) + c$$

$$= \frac{8}{x} + c$$

where c is an arbitrary constant.

Since $v = 5$ when $x = 1$:

$$\frac{1}{2}(5)^2 = \frac{8}{1} + c$$

So:
$$c = 4\tfrac{1}{2}$$

and:
$$\frac{1}{2}v^2 = \frac{8}{x} + 4\tfrac{1}{2}$$

or:
$$v^2 = \frac{16}{x} + 9$$

As x tends to infinity, $\dfrac{16}{x}$ tends to zero and so $v^2 \to 9$.

Hence $v \to 3$, since v is initially positive.

Exercise 1B

1 A particle P moves along the positive x-axis with constant acceleration f m s^{-2} ($f > 0$). When P is at the origin its velocity is u m s^{-1} in the direction Ox. Find the velocity v m s^{-1} in terms of the displacement x metres from O.

2 A particle P moves along a straight line. Its acceleration when it has a displacement of x metres from O, a fixed point in the line, is of magnitude $4x \, \mathrm{m \, s^{-2}}$ and directed towards O. Given that the velocity of P is zero when $x = 5$, find the speed of P when $x = 3$.

3 A particle P moves along a straight line and when its distance from a fixed point O is x metres, its retardation is $(20 + 2x) \, \mathrm{m \, s^{-2}}$. Given that the speed of P is $40 \, \mathrm{m \, s^{-1}}$ when $x = 0$, obtain an expression for v in terms of x. Find the value of x when P comes to rest.

4 The acceleration of a particle P, moving on a straight line, is $\lambda x^2 \, \mathrm{m \, s^{-2}}$ $(\lambda > 0)$ when P has a displacement x metres from O, a fixed point in the line. The acceleration is towards O. The speed of P is zero when $x = a$. Find the speed of P when $x = 0$.

5 When a particle P has a displacement of x metres from a fixed point O in a straight line, its acceleration is $\dfrac{k}{x} \mathrm{m \, s^{-2}}$ $(k > 0)$ away from O. It has zero velocity when $x = 1$. Find x in terms of v and k.

6 The acceleration of a particle P, moving in a straight line, is $4x^3 \, \mathrm{m \, s^{-2}}$ when P has a displacement x metres from O, a fixed point in the line. The acceleration is away from O. Given that $v = 2$ when $x = 0$, obtain the value of x when $v = 6$.

7 A particle P moves along the positive x-axis. Its acceleration is $(x + 3) \, \mathrm{m \, s^{-2}}$ when its displacement from the origin O is x metres. Given that initially, when $t = 0$, the velocity of P is $3 \, \mathrm{m \, s^{-1}}$ in the direction Ox and $x = 0$, obtain (a) the speed $v \, \mathrm{m \, s^{-1}}$ of P as a function of x (b) x as a function of t.

8 A particle moves along the positive x-axis. When it has a displacement of x metres from O its acceleration is of magnitude $\dfrac{4}{x^3} \mathrm{m \, s^{-2}}$ and directed towards O. Given that initially, when $t = 0$, the velocity of the particle is $2 \, \mathrm{m \, s^{-1}}$ in the direction Ox and $x = 1$, obtain (a) v as a function of x (b) x as a function of t.

9   A particle P moves along the positive x-axis and its
    acceleration is $(e^{2x})\,\mathrm{m\,s^{-2}}$ when its displacement from O is
    x metres. Given that initially, when $t = 0$, the velocity of P
    is $u\,\mathrm{m\,s^{-1}}$ towards O and $x = \ln u$, obtain (a) v as a
    function of x (b) x as a function of t.

10 A particle moves on the positive x-axis. When its displacement
    from O is x metres, its acceleration is of magnitude $\dfrac{1}{x^3}\,\mathrm{m\,s^{-2}}$
    and directed towards O. Given that initially, when $t = 0$, the
    particle is at rest with $x = 1$, find the time the particle takes to
    reach $x = \frac{1}{4}$.

11 A particle P moves on the positive x-axis and its acceleration,
    when it has a displacement of x metres from O, is
    $(4 \sin x)\,\mathrm{m\,s^{-2}}$ in the direction Ox. Given that when $x = 0$ the
    velocity of P is $6\,\mathrm{m\,s^{-1}}$ in the direction Ox, find the maximum
    speed of P.

12 A particle P moves on the positive x-axis and its acceleration,
    when it has a displacement of x metres from O, is $(e^{2x})\,\mathrm{m\,s^{-2}}$.
    Given that initially, when $t = 0$, the speed of P is $1\,\mathrm{m\,s^{-1}}$ in
    the direction Ox and $x = 0$, obtain x as a function of t and
    show that $x \to \infty$ as $t \to 1$.

13 A particle P moves on the positive x-axis. When it has a
    displacement of x metres from O, it has acceleration
    $3\sqrt{x}\,\mathrm{m\,s^{-2}}$ in the direction Ox. When $t = 0$ the particle starts
    from rest at O. Show that $x = \frac{1}{16}t^4$.

14 A particle P moves on the positive x-axis and its acceleration,
    when it has a displacement of x metres from O, is
    $2 \sin^2 x\,\mathrm{m\,s^{-2}}$. When $x = \dfrac{\pi}{4}$ the velocity of P is $1\,\mathrm{m\,s^{-1}}$ in the
    direction Ox. Find the velocity of P when $x = \dfrac{\pi}{3}$.

15 A particle P is moving along the positive x-axis. At time t
    seconds the particle is at a distance x metres from the origin O,
    its speed is $v\,\mathrm{m\,s^{-1}}$ and its acceleration is $(x + 2)\,\mathrm{m\,s^{-1}}$. Given
    that when $t = 0$, the velocity of P is $2\,\mathrm{m\,s^{-1}}$ in the direction
    Ox, and $x = 0$,
    (a) show that $v = x + 2$
    (b) find x as a function of t.

SUMMARY OF KEY POINTS

1 For a particle P travelling in a straight line, which at time t seconds has a displacement x metres from a fixed point O of the line, the acceleration $a \, \mathrm{m\,s}^{-2}$ and the velocity $v \, \mathrm{m\,s}^{-1}$ as functions of time are given by:

$$a = \frac{\mathrm{d}v}{\mathrm{d}t}$$

$$v = \frac{\mathrm{d}x}{\mathrm{d}t}$$

$$a = \frac{\mathrm{d}^2 x}{\mathrm{d}t^2}$$

2 When the acceleration is a function of the displacement then:

$$a = v\frac{\mathrm{d}v}{\mathrm{d}x} = \frac{\mathrm{d}}{\mathrm{d}x}\left(\frac{1}{2}v^2\right)$$

Elastic springs and strings

In the work considered so far, all strings have been assumed to be inextensible. In fact no string is totally inextensible although many can be assumed to be so. Those strings which can be stretched are called **elastic**.

An **elastic string** is assumed to regain its original or **natural length** when the force causing it to stretch is removed. The extra length resulting from the force acting on an elastic string is called the **extension**. A stretched string will have a **tension**.

Consider an elastic string AB which has end A fixed. When there is no force applied to end B the length of the string is its natural length l and there is no tension in the string. However, when a force $\mathbf{F}$ is applied to the end B, the string will stretch.

The length of the string is now $l + x$ where x is the extension. The tension $\mathbf{T}$ in the string will be equal in magnitude to the extending force $\mathbf{F}$.

An elastic spring behaves in the same manner as an elastic string when stretched. However, a spring can also be compressed.

With end A fixed and a force $\mathbf{F}$ applied to end B in the direction BA the spring will be compressed.

The spring will exert a **thrust** $\mathbf{T}$ of magnitude equal to the magnitude of the compressing force $\mathbf{F}$.

If an elastic string is overstretched it may break or become permanently deformed. Similarly if a spring is overstretched or overcompressed it may become deformed and not return to its natural length when released. The point at which permanent deformation takes place is called the **elastic limit**. It will be assumed in what follows that all strings and springs do not reach this limit.

2.1 Hooke's law

Consider a simple experiment in which bodies of different masses are attached to the free end of an elastic string, the other end of which is fixed. The resulting extensions of the string may be recorded.

Since the body is in equilibrium $T = mg$ where T is the tension in the string and m is the mass of the body.

The results of such an experiment show that:

the tension in the string is proportional to the extension produced.

This is known as **Hooke's law**.

Hooke's law is normally used in the form:

$$T = \frac{\lambda x}{l}$$

where T is the tension in the string, x is the extension produced and l is the natural length of the string. The constant of proportionality, λ, is called the **modulus of elasticity** or simply the **modulus** of the string, and depends on the nature of the string.

You must remember that all strings are capable of being stretched to some extent although they are often modelled as though this were not the case. Also wires can be stretched and consequently in some situations they are modelled as elastic strings. Most wires would stretch by only a small amount even if the applied force were of large magnitude.

Consideration of the equation:

$$T = \frac{\lambda x}{l}$$

will show you that if a large numerical value for T only produces a small extension x then λ must also be large. The table gives the modulus of elasticity for wires of diameter 1 mm of some materials. Units must balance in any equation. Both x and l are lengths and consequently are measured in the same units, so λ must have the same units as force. For a force in newtons, λ is also in newtons.

Material	Modulus of elasticity
Aluminium	5.5×10^4
Copper	8.8×10^4
Brass	7.3×10^4
Mild steel	1.6×10^5

Hooke's law applies equally to springs which are compressed with the thrust being proportional to the compression.

Example 1

An elastic string has natural length 2 m and modulus 12 N. Calculate (a) the tension in the string when the extension is 0.2 m (b) the extension of the string when the tension is 8 N.

(a) Using Hooke's law:

$$T = \frac{\lambda x}{l}$$

with $\lambda = 12\,\text{N}$, $l = 2\,\text{m}$, $x = 0.2\,\text{m}$ gives:

$$T = \frac{12 \times 0.2}{2} = 1.2$$

The tension is 1.2 N.

(b) Using Hooke's law:

$$T = \frac{\lambda x}{l}$$

with $\lambda = 12\,\text{N}$, $T = 8\,\text{N}$, $l = 2\,\text{m}$ gives:

$$8 = \frac{12x}{2}$$

$$x = \frac{2 \times 8}{12} = 1.333$$

The extension is 1.33 m.

Example 2

A spring has modulus 30 N and natural length 0.5 m. Calculate the thrust in the spring when it is compressed a distance of 10 cm.

Let the thrust be T N.

Compression $= 10\,\text{cm} = 0.1\,\text{m}$, natural length $= 0.5\,\text{m}$, $\lambda = 30\,\text{N}$

Using Hooke's law:

$$T = \frac{\lambda x}{l}$$

gives:

$$T = 30 \times \frac{0.1}{0.5}$$

$$T = 6$$

The thrust is 6 N.

Example 3

A and B are two fixed points on a horizontal table with $AB = 5l$. A particle P is attached to A by means of a light elastic string of modulus λ and natural length l and to B by means of a second light elastic string of modulus 2λ and natural length $2l$. The particle rests in equilibrium at the point O. Find the distance OA.

$$A \xrightarrow{\hspace{2cm} x \hspace{2cm}} \underset{\substack{T_1 \quad P \quad T_2}}{\overset{O}{\bullet}} \xrightarrow{\hspace{1.5cm} 5l - x \hspace{1.5cm}} B$$

Let $OA = x$; then $OB = 5l - x$.

Let the tension in OA be T_1 and the tension in OB be T_2.

Using Hooke's law:
$$T = \frac{\lambda x}{l}$$

gives:
$$T_1 = \frac{\lambda(x - l)}{l} \qquad T_2 = \frac{2\lambda[(5l - x) - 2l]}{2l}$$

As P is in equilibrium:
$$T_1 = T_2$$

So:
$$\frac{\lambda(x - l)}{l} = \frac{2\lambda[(5l - x) - 2l]}{2l}$$
$$x - l = 3l - x$$
$$2x = 4l$$
$$x = 2l$$

The distance OA is $2l$.

Equilibrium of a particle suspended on an elastic string or spring

Consider a string or spring with one end fixed to a ceiling and a particle attached to the free end. When the particle is hanging in equilibrium the tension T in the string can be found by resolving vertically, giving:
$$T = mg$$

Using Hooke's law gives:
$$T = \frac{\lambda x}{l}$$

where x is the extension in the string and l is its natural length.

So:
$$mg = \frac{\lambda x}{l}$$

or the extension is given by:
$$x = \frac{mgl}{\lambda}$$

Example 4

A light elastic string of natural length 50 cm and modulus 40 N has one end fixed. A particle of mass 2 kg is attached to the other end and hangs in equilibrium. Calculate the extension in the string.

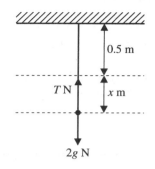

Let the extension be x m.

The particle is in equilibrium so resolving vertically gives:

$$T = 2g$$

Using Hooke's law, $T = \dfrac{\lambda x}{l}$ with $\lambda = 40$ N, $l = 50$ cm $= 0.5$ m and $T = 2g$ N gives:

$$2g = 40 \times \frac{x}{0.5}$$

$$x = \frac{g}{40} = 0.245$$

The extension in the string is 0.245 m or 24.5 cm.

Example 5

A rope is tied to a branch of a tree. When the rope is hanging freely its length is 2.5 m. When a child of mass 40 kg hangs in equilibrium from the end of the rope, the length of the rope is 2.7 m. Stating clearly any assumptions made, choose suitable models for the rope and the child and hence calculate the modulus of the rope.

The mass of the rope will be small compared with that of the child. The rope can therefore be assumed to be light. It stretches under the weight of the child, so it can be modelled as a light elastic string of natural length 2.5 m.

The child is hanging on to the end of the rope so the contact between the child and the rope is restricted to this point. The child can therefore be modelled as a particle.

Let the tension in the rope be T N.

As the child is in equilibrium, resolving vertically gives:

$$T = 40g \qquad (1)$$

The extension in the rope is $(2.7 - 2.5) = 0.2\,\text{m}$.

Using Hooke's law, $T = \dfrac{\lambda x}{l}$, for the rope with $l = 2.5\,\text{m}$ and $x = 0.2\,\text{m}$ gives:

$$T = \lambda \times \frac{0.2}{2.5} \qquad (2)$$

Eliminating T between equations (1) and (2) gives:

$$\lambda \times \frac{0.2}{2.5} = 40 \times 9.8$$

$$\lambda = 40 \times 9.8 \times \frac{2.5}{0.2}$$

$$\lambda = 4900$$

The modulus of the rope is $4900\,\text{N}$.

Example 6

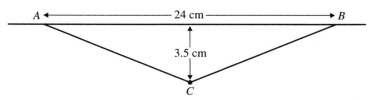

The diagram shows two fixed points A and B on the same horizontal ceiling where $AB = 24\,\text{cm}$. A light elastic string has one end attached to A and the other end attached to B. The natural length of the string is $20\,\text{cm}$. A particle of mass $0.4\,\text{kg}$ is attached to the mid-point of the string and hangs in equilibrium at a point C which is $3.5\,\text{cm}$ below the mid-point of the line AB. Calculate the modulus of the string.

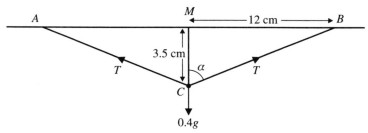

Let M be the mid-point of AB; then $MB = 12\,\text{cm}$.

Using Pythagoras' theorem:

$$BC^2 = 12^2 + 3.5^2$$

$$= 144 + 12.25$$

$$= 156.25$$

So: $$BC = 12.5\,\text{cm}$$

and:
$$\cos \alpha = \frac{3.5}{12.5} = \frac{7}{25}$$

Resolving vertically gives:
$$2T \cos \alpha = 0.4g \qquad (1)$$

Using Hooke's law for BC:
$$T = \lambda \frac{(12.5 - 10)}{10} = \frac{\lambda \times 2.5}{10} = \frac{\lambda}{4} \qquad (2)$$

Eliminating T from equations (1) and (2) gives:
$$\frac{2\lambda}{4} \times \frac{7}{25} = 0.4g$$

$$\lambda = \frac{25 \times 4 \times 0.4g}{7 \times 2} = 28\,\text{N}$$

The modulus of the string is 28 N.

Example 7

Two light elastic strings AB and BC are fastened together at B. String AB has natural length 1 m and modulus of elasticity 15 N and string BC has natural length 0.5 m and modulus of elasticity 18 N. A is attached to a point on a ceiling. A particle of mass 0.5 kg is attached at C and the system hangs in equilibrium. Find the total length of the two strings.

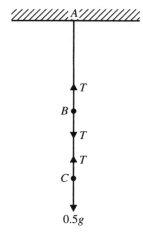

Consider the point B where the strings are joined. The only forces acting at this point are the tensions in the two strings. As the join is in equilibrium, the two tensions must be equal.

Let the tensions in both strings be T.

Let the extension of AB be x m and the extension of BC be y m.

Resolving vertically for the particle gives:

$$T = 0.5g \tag{1}$$

Using Hooke's law for string AB:

$$T = \frac{15x}{1} \tag{2}$$

Using Hooke's law for string BC:

$$T = \frac{18y}{0.5} \tag{3}$$

Eliminating T between equations (1) and (2) gives:

$$15x = 0.5g$$

So:

$$x = \frac{0.5g}{15}$$

Eliminating T between equations (1) and (3) gives:

$$\frac{18y}{0.5} = 0.5g$$

So:

$$y = \frac{0.5g}{36}$$

Hence the total length $= 1.5 + x + y = 1.5 + \dfrac{0.5g}{15} + \dfrac{0.5g}{36} = 1.96 \, \text{m}.$

Motion of a particle attached to the end of an elastic string or spring

Consider now the motion of a particle attached to one end of an elastic string which is hanging vertically and has its other end fixed. The forces acting on the particle are gravity and the tension in the string. If the particle is at some point other than its equilibrium position there will be a resultant force acting on it and this will produce an acceleration.

When the particle P is in equilibrium, using Hooke's law gives:

$$T = \frac{\lambda e}{l}$$

where e is the extension.

Hence:

$$e = \frac{Tl}{\lambda}$$

or, as $T = mg$:

$$e = \frac{mgl}{\lambda}$$

When the particle P is at some point where the extension, x, is greater than $\dfrac{mgl}{\lambda}$ then the tension in the string will be greater than the weight of the particle and the particle will move with an acceleration a.

Using Hooke's law in this position gives $T = \dfrac{\lambda x}{l}$.

Using the equation of motion, $F = ma$, gives:

$$T - mg = ma$$

So:

$$\frac{\lambda x}{l} - mg = ma$$

or:

$$a = \frac{\lambda x}{ml} - g$$

The resulting acceleration depends on the extension in the string and is constantly changing.

Example 8

A particle of mass 1 kg is attached to one end A of a light elastic string AB of modulus 25 N and natural length 0.5 m. The other end B of the string is fixed and A is held at a distance of 0.75 m vertically below B and then released. Calculate the initial acceleration of A.

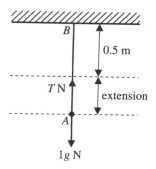

extension of the string is $(0.75 - 0.5)\,\text{m} = 0.25\,\text{m}$.

he initial acceleration of A be $a\,\text{m s}^{-2}$.

the equation of motion $F = ma$ gives:

$$T - 1g = a \qquad\qquad (1)$$

Hooke's law: $\qquad T = \dfrac{\lambda x}{l}$

Substituting $\lambda = 25\,\text{N}, \quad x = (0.75 - 0.5)\,\text{m} = 0.25\,\text{m}, \quad l = 0.5\,\text{m}$ gives:

$$T = 25 \times \frac{0.25}{0.5}$$

$$T = 12.5$$

Substituting $T = 12.5$ in equation (1) gives:

$$12.5 - g = a$$

$$a = 12.5 - 9.8 = 2.7$$

The initial acceleration of the particle is $2.7\,\text{m s}^{-2}$.

Example 9

A particle of mass $0.5\,\text{kg}$ is attached to one end A of a light elastic string AB of modulus $19.6\,\text{N}$ and natural length $1\,\text{m}$. The end B of the string is attached to a point O on a plane inclined at an angle $\arctan\left(\frac{3}{4}\right)$ to the horizontal. The particle is held at rest on the slope at a point that is $1.5\,\text{m}$ from O down a line of greatest slope of the plane. The particle is released from rest and moves up the slope. Find its initial acceleration
(a) when the plane is assumed smooth,
(b) when the plane is rough with coefficient of friction 0.25.

Let the tension in the string be $T\,\text{N}$.

Using Hooke's law gives: $\qquad T = \dfrac{19.6(1.5 - 1)}{1} = 9.8\,\text{N}$

(a) When the plane is assumed smooth:

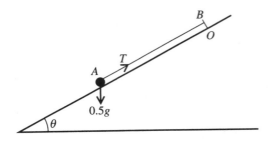

Using $F = ma$ parallel to the plane gives: $\quad T - 0.5g \sin\theta = 0.5a$

As $\tan\theta = \frac{3}{4}$ it follows that $\sin\theta = \frac{3}{5}$.

So: $\qquad a = \left(\dfrac{9.8 - 0.5 \times \frac{3}{5}g}{0.5}\right) = 13.7\,\text{m s}^{-2}$

(b) When the plane is assumed rough:

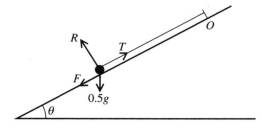

Resolving perpendicular to the plane gives:
$$R = 0.5g\cos\theta = 0.5g \times \tfrac{4}{5}$$
as $\cos\theta = \tfrac{4}{5}$.

Hence the friction force $F = \mu R$ is $0.25 \times 0.5g \times \tfrac{4}{5} = 0.1g$.

Using $F = ma$ parallel to the plane gives:
$$T - 0.5g\sin\theta - \mu R = 0.5a$$
$$T - 0.3g - 0.1g = 0.5a$$
$$9.8 - 0.3g - 0.1g = 0.5a$$
$$a = 2(9.8 - 0.3g - 0.1g) = 1.2g = 11.8\,\mathrm{m\,s}^{-2}$$
So the acceleration is $11.8\,\mathrm{m\,s}^{-2}$.

Exercise 2A

Whenever a numerical value of g is required, take $g = 9.8\,\mathrm{m\,s}^{-2}$.

1 An elastic string has natural length 2 m and modulus 30 N. Find the tension in the string when the extension is 0.5 m.

2 An elastic string has modulus 25 N and the tension in the string is 40 N. Find the natural length if the extension is 0.1 m.

3 The tension in an elastic string of modulus 30 N is 50 N. The length of the string is 1.5 m. Calculate the natural length of the string.

4 A spring of modulus 40 N and natural length 0.5 m is compressed to a length of 0.4 m. Calculate the thrust produced.

5 A spring of natural length 1 m exerts a thrust of 45 N when compressed to a length of 0.75 m. Calculate the modulus of the spring.

6 A force of 5 N is applied to an elastic string of natural length 2.5 m and modulus 4 N. Calculate the extension produced.

7 An elastic spring of modulus 50 N has one end fixed. When a particle of mass 2 kg is hanging in equilibrium attached to the free end of the spring, the spring is extended by 5 cm. Find the natural length of the spring.

8 An elastic string of modulus 20 N has one end fixed. When a particle of mass 1 kg is hanging in equilibrium from the free end of the string, the length of the string is 1.4 m. Determine the natural length of the string.
The particle of mass 1 kg is removed and replaced by a particle of mass 0.75 kg. Calculate the new length of the string.

9 A particle of mass 6 kg hangs vertically from the free end A of an elastic string AB. End B of the string is fixed. If the extension in the string is equal to its natural length, determine the modulus of the string.

10 Two elastic strings, AB of modulus 15 N and natural length 1 m and CD of modulus 18 N and natural length 0.5 m are joined, B to C, to form one long string. End A is fixed to a ceiling. A particle of mass 0.5 kg is attached to D and hangs in equilibrium. Calculate the length of the combined string.

11 Two elastic strings both of natural length 1.5 m and moduli 8 N and 6 N respectively are joined together at both ends. One end of the combined string is fixed and a particle of mass 2 kg is attached to the other end and hangs in equilibrium with the string vertical. Calculate the tension in each string.

12 A and B are two fixed points on a smooth horizontal table with $AB = 1.5$ m. A particle P is in equilibrium, attached to A by means of a light elastic string of natural length 0.5 m and modulus of elasticity λ N and to B by means of a light elastic string of natural length 0.75 m and modulus of elasticity 2λ N. Given that the tension in AP is 10 N, find the value of λ.

13 A and B are two fixed points on a smooth horizontal table. A particle P is in equilibrium, attached to A by means of a light elastic string of natural length l and modulus of elasticity λ and to B by means of a light elastic string of natural length $3l$ and modulus of elasticity λ. Given that the extension in the string AP is $\frac{1}{2}l$, find the distance AB.

14 A particle P of mass 2 kg is attached to the mid-point of a
    light elastic string of natural length 2 m and modulus of
    elasticity 20 N. One end of the string is attached to point A of
    a ceiling and the other end is attached to point B on the floor
    3 m vertically below A. Calculate the tension in AP and the
    tension in PB.

15 A ball of mass 0.5 kg is attached to one end of a cord of
    unstretched length 0.6 m whose other end is fixed. When a
    horizontal force of magnitude Q N is applied to the ball,
    holding the ball in equilibrium, the cord increases in length by
    10% and is inclined at an angle of arcsin $\frac{3}{5}$ to the downward
    vertical. Choose suitable models for the ball and cord stating
    clearly the assumptions you make. Hence calculate the value
    of Q and the modulus of the cord.

16 A particle of mass 2 kg is attached to one end of a light elastic
    string of natural length 0.5 m and modulus 60 N whose other
    end is fixed. When the particle is pulled aside by a horizontal
    force S and held in equilibrium the extension in the string is
    0.2 m. Calculate the angle the string makes with the downward
    vertical and the magnitude of S.

17 A particle of mass 4 kg is attached to one end A of a light
    elastic string AB of modulus 20 N and natural length 0.8 m.
    The end B of the string is attached to a point on a smooth
    plane inclined at an angle arcsin $\frac{5}{13}$ to the horizontal. The
    particle rests in equilibrium on the plane with AB along a line
    of greatest slope. Calculate (a) the tension in the string (b) the
    extension of the string.

18 A particle of mass 0.8 kg is attached to one end of a light
    elastic string AB of modulus of elasticity 14.7 N and natural
    length 1.5 m. The end B of the string is attached to a point O
    on a plane inclined at an angle α to the horizontal, where
    $\sin \alpha = \frac{5}{12}$. The particle is held at rest on the slope at a point
    that is 2 m below O down a line of greatest slope. The particle
    is released from rest and moves up the slope which is assumed
    to be smooth. Find its initial acceleration.

19 A particle of mass 0.3 kg is attached to one end of a light
    elastic string AB of modulus of elasticity 4.9 N and natural

length 0.5 m. The end B of the string is attached to a point O on a plane inclined at an angle α to the horizontal, where $\sin \alpha = \frac{3}{5}$. The particle is held at rest on the slope at a point that is 1 m below O down a line of greatest slope. The particle is released from rest and moves up the slope which is assumed to be rough with coefficient of friction $\frac{1}{4}$. Find its initial acceleration.

20 A metal sphere of mass 3 kg is attached to one end A of a spring AB of modulus 90 N and natural length 0.5 m. End B of the spring is fixed to a point on a surface which is inclined at an angle $\arcsin \frac{3}{5}$ to the horizontal. The sphere rests in equilibrium on the surface with AB along a line of greatest slope and A below B. A horizontal force of magnitude Q N is applied to the sphere causing the spring to be compressed to a length 0.4 m.
(a) Assuming that the surface can be modelled as a smooth plane, choose suitable models for the sphere and spring and hence calculate the value of Q.
(b) Refining your model of the surface to a rough plane and given that the sphere is on the point of moving down the plane when $Q = 35$, calculate the coefficient of friction between the surface and the sphere.

21 A particle of mass 2 kg is attached to one end A of a light spring AB of modulus 60 N and natural length 0.5 m. The spring rests on a smooth horizontal table with one end fixed. The particle is held at rest with AB in a straight line. The particle is then released. Calculate its initial acceleration (a) if AB is initially 0.7 m (b) if AB is initially 0.3 m.

22 A particle P of mass m is suspended from two strings AP and BP where A and B are attached to two points on a horizontal ceiling distance $\frac{5l}{4}$ apart as shown in the diagram.
AP is inelastic and of length $\frac{3l}{4}$. BP is elastic and of length l.
The modulus of elasticity of BP is λ. Show that the natural length L of BP is given by:

$$L = \frac{5\lambda l}{3mg + 5\lambda}$$

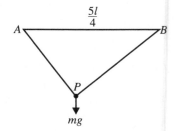

23 A particle of mass 0.5 kg is attached to one end A of a light elastic string AB of modulus 40 N and natural length 0.8 m. The other end B of the string is fixed and A is initially hanging in equilibrium with the string vertical. A is then pulled vertically downwards a further 0.1 m and released. Calculate the initial acceleration of the particle.

2.2 Energy stored in an elastic string or spring

Consider a particle attached to one end of an elastic string whose other end is fixed on a smooth horizontal table. If the string is stretched beyond its natural length by pulling the particle along the table and then the particle is released, the particle will be seen to move on the table. That is, it will acquire kinetic energy. As the motion is horizontal there can be no change in the gravitational potential energy so it follows that when stretched, the string has energy stored in it. This form of potential energy is called the **elastic potential energy** (E.P.E.) of the string.

By the work–energy principle this elastic potential energy must be equal to the work done in stretching the string, that is the work done against the tension in the string. By Hooke's law, the tension varies with the extension in the string, $T = \dfrac{\lambda x}{l}$. A graph of the tension against the extension is a straight line:

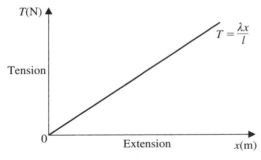

Hence it is not possible to calculate the work done against the tension by simply multiplying the tension by the extension. Instead an integration method, similar to that used to find the area under a graph (as in Book P1, Chapter 6) must be used.

First consider the work done in extending the string from an extension s to an extension $s + \delta s$ where δs is a very small length. The tension may be considered to be constant over this very small distance.

Let the tension when the extension is s be T_s.

Using: work done = force × distance moved

Gives: work done = $T_s \times \delta s$

From Hooke's law: $T_s = \dfrac{\lambda s}{l}$

and so: work done = $\left(\dfrac{\lambda s}{l}\right) \delta s$

The total work done in stretching the string from its natural length (where $s = 0$) to a length $l + x$ (where $s = x$) is given by adding the individual quantities of work for each extension δs together.

Hence: work done = $\displaystyle\sum_{s=0}^{x} \dfrac{\lambda s \delta s}{l}$

Now let δs become even smaller, that is let $\delta s \to 0$.

$$\text{Work done} = \lim_{\delta s \to 0} \sum_{s=0}^{x} \frac{\lambda s \delta s}{l}$$

$$= \int_{0}^{x} \frac{\lambda s}{l}\, \mathrm{d}s$$

$$= \left[\frac{\lambda s^2}{2l}\right]_{0}^{x}$$

$$= \frac{\lambda x^2}{2l}$$

■ **The work done in stretching an elastic spring with modulus λ from its natural length l to a length $(l + x)$ is $\dfrac{\lambda x^2}{2l}$**

When λ is in newtons and x and l are in metres, the work done is in joules.

By the work–energy principle, the E.P.E. stored in a string extended by a length x beyond its natural length l is equal to the work done in stretching the string, that is $\dfrac{\lambda x^2}{2l}$

The elastic potential energy stored in a spring compressed by a length x from its natural length l is also given by $\dfrac{\lambda x^2}{2l}$

A string (or spring) which has its extension (or compression) increased from x_1 to x_2 will have its elastic energy increased. This increase is given by:

increase in E.P.E. = final E.P.E. − initial E.P.E.

$$= \frac{\lambda x_2^2}{2l} - \frac{\lambda x_1^2}{2l}$$

Example 10

An elastic string has natural length 2 m and modulus 15 N. Calculate the energy stored in the string when the string is stretched to a length of 2.6 m.

The energy stored in the string (E.P.E.) $= \dfrac{\lambda x^2}{2l}$

Since $\lambda = 15\,\text{N}$, $l = 2\,\text{m}$ and $x = 2.6 - 2 = 0.6\,\text{m}$

$$\text{E.P.E.} = \frac{15 \times 0.6^2}{2 \times 2} = 1.35$$

The energy stored in the string is 1.35 J.

Example 11

A spring of natural length 0.75 m and modulus 80 N is initially compressed to a length of 0.7 m. It is then further compressed to a length of 0.5 m. Calculate the increase in the energy in J stored in the spring.

$$\text{Initial compression } x_1 = 0.75 - 0.7 = 0.05\,\text{m}$$

$$\text{Final compression } x_2 = 0.75 - 0.5 = 0.25\,\text{m}$$

Using: $\quad\quad$ increase in energy $= \dfrac{\lambda x_2^2}{2l} - \dfrac{\lambda x_1^2}{2l}$

with $\lambda = 80\,\text{N}$, $l = 0.75\,\text{m}$, $x_1 = 0.05\,\text{m}$ and $x_2 = 0.25\,\text{m}$ gives:

$$\text{increase} = \frac{80 \times 0.25^2}{2 \times 0.75} - \frac{80 \times 0.05^2}{2 \times 0.75}$$
$$= 3.2$$

The increase in the energy stored in the spring is 3.2 J.

Exercise 2B

Whenever a numerical value of g is required, take $g = 9.8\,\text{m}\,\text{s}^{-2}$.

1   An elastic string has natural length 0.8 m and modulus 10 N. Calculate the elastic potential energy stored in the string when the string is stretched to a length of 1.2 m.

2   An elastic spring of natural length 0.5 m and modulus 150 N is compressed to a length of 0.4 m. Calculate the elastic potential energy stored in the spring.

3   An elastic string has natural length 0.5 m and modulus 8 N. Calculate the work done in stretching the string (a) from its natural length to a length of 0.8 m (b) from a length of 0.75 m to 1.3 m.

4   An elastic spring has natural length 0.8 m and modulus 200 N. Calculate the work done in compressing the spring (a) from its natural length to a length of 0.7 m (b) from a length of 0.75 m to a length of 0.65 m.
In each of parts (a) and (b) the spring is compressed by the same distance. What do you notice about the changes in energy?

5   A particle of mass 1.5 kg is hanging in equilibrium at the free end of a light elastic string of natural length 2 m and modulus 16 N. The other end of the string is fixed. The particle is pulled downwards until the string has length 4 m. Calculate the increase in the elastic energy stored in the string.

6   An elastic string has natural length l. One end of the string is fixed and a particle of mass m is attached to the other end. When the particle hangs in equilibrium the length of the string is $2l$. Calculate the energy stored in the string.

7   An elastic string has natural length l. One end of the string is fixed and a particle of mass $2m$ is attached to the other end. When the particle hangs in equilibrium the length of the string is $\frac{3}{2}l$. Calculate the work done in moving the particle from its equilibrium position to the point where the string has length $2l$.

8   One end of an elastic string of natural length l is fixed. When a particle of mass $2m$ hangs in equilibrium from the free end the length of the string is $2l$. Calculate the energy stored in the string when a particle of mass m hangs in equilibrium from the free end of the string.

2.3 Problems involving kinetic energy, gravitational potential energy and elastic potential energy

In Book M2 the work–energy principle was applied to problems involving kinetic energy and gravitational potential energy. This work can now be extended to include problems involving elastic strings or springs and elastic potential energy.

If a particle attached to the free end of an elastic string is held in equilibrium with the elastic string stretched and then released, the particle will move. As the extension of the string becomes less, the elastic potential energy stored in the string will also be reduced. The elastic potential energy is transferred to the particle which shows a corresponding increase in kinetic energy. If there is any vertical motion, the gravitational potential energy must also be taken into account.

By the work–energy principle, the total change of the kinetic, gravitational and elastic potential energies of the particle must be equal to the work done by any external forces acting on the particle.

Example 12

A particle P of mass 2 kg is attached to one end of a light elastic string of natural length 1.6 m and modulus 20 N. The other end of the string is fixed to a point O on a smooth horizontal table. P is released from rest at a point on the table 2.5 m from O and moves on the table. Calculate the speed of P when the string becomes slack.

The string becomes slack when it returns to its natural length, that is when $OP = 1.6$ m.

Let the speed of P when $OP = 1.6$ m be v m s^{-1}.

When the string is slack it has no elastic potential energy.

The energy stored in the string when $OP = 2.5$ m is given by:

$$\text{elastic potential energy} = \frac{\lambda x^2}{2l}$$

where $\lambda = 20$ N, $l = 1.6$ m, $x = (2.5 - 1.6) = 0.9$ m

So:
$$\text{E.P.E.} = \frac{20 \times 0.9^2}{2 \times 1.6}\,\text{J}$$

$$\text{Initial kinetic energy} = 0\,\text{J}$$

$$\text{Final kinetic energy} = \tfrac{1}{2}mv^2 = \tfrac{1}{2} \times 2 \times v^2 = v^2\,\text{J}$$

By the work–energy principle:

$$\text{kinetic energy gained by particle} = \text{E.P.E. lost by string}$$

So:
$$v^2 = \frac{20 \times 0.9^2}{2 \times 1.6}$$

$$v = 2.25$$

The speed of P when the string becomes slack is 2.25 m s^{-1}.

Example 13

A particle of mass 0.5 kg is attached to one end A of a light elastic string of natural length 1 m and modulus 5 N. The other end B of the string is fixed to a point on a ceiling. The particle is held at B and then released. Calculate

(a) the length of the string when the particle reaches its lowest point
(b) the speed of the particle when it passes through the equilibrium position.

(a)

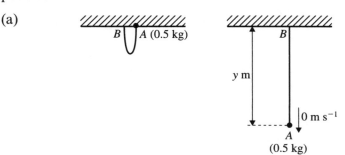

Let the length of the string when the particle is at its lowest point be y m. The extension will then be $(y-1)$ m. At this lowest point the particle comes instantaneously to rest, that is its speed at its lowest point is $0\,\mathrm{m\,s^{-1}}$.

Using:
$$\text{E.P.E.} = \frac{\lambda x^2}{2l}$$

with $\lambda = 5\,\mathrm{N}$, $l = 1\,\mathrm{m}$ and $x = (y-1)\,\mathrm{m}$ gives:

$$\text{E.P.E.} = \frac{5(y-1)^2}{2\times 1}\ \mathrm{J}$$

The gravitational P.E. lost by the particle $= mgh$
$$= 0.5 \times 9.8 \times y\ \mathrm{J}$$

Initial K.E. of particle $=$ final K.E. of particle $= 0\,\mathrm{J}$

Hence by the work–energy principle:

E.P.E. gained by string $=$ gravitational P.E. lost by particle

$$\frac{5(y-1)^2}{2\times 1} = 0.5 \times 9.8 \times y$$
$$5(y-1)^2 = 9.8 \times y$$
$$5(y^2 - 2y + 1) - 9.8y = 0$$
$$5y^2 - 19.8y + 5 = 0$$
$$y = \frac{(19.8 \pm \sqrt{(19.8^2 - 4 \times 5 \times 5)})}{2 \times 5}$$
$$y = 0.27 \text{ or } 3.69$$

But y must be greater than 1 m, the natural length of the string.

Hence the length of the string when the particle reaches its lowest point is 3.69 m.

(b) First the equilibrium position must be found.
When the particle is hanging in equilibrium, resolving vertically gives:

$$T = 0.5g$$

Using Hooke's law: $$T = \frac{\lambda x}{l}$$

with $\lambda = 5\,\text{N}$ and $l = 1\,\text{m}$ gives:

$$T = \frac{5x}{1}$$

So: $$0.5g = \frac{5x}{1}$$

$$x = \frac{0.5 \times 9.8}{5} = 0.98$$

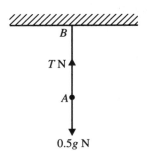

The extension in the equilibrium position is 0.98 m and so the length AB is 1.98 m.

Consider the particle falling from rest to its equilibrium position. Let the speed of the particle when AB is 1.98 m be $v\,\text{m}\,\text{s}^{-1}$.

Gravitational P.E. lost by particle $= mgh = 0.5 \times 9.8 \times 1.98\,\text{J}$

K.E. gained by particle $= \tfrac{1}{2}mv^2 = \tfrac{1}{2} \times 0.5 \times v^2\,\text{J}$

E.P.E. of string at the equilibrium position $= \dfrac{\lambda x^2}{2l} = \dfrac{5 \times 0.98^2}{2 \times 1}\,\text{J}$

By the work–energy principle:

$$(\text{K.E.} + \text{E.P.E.}) \text{ gained} = \text{P.E. lost}$$

$$\frac{1}{2} \times 0.5 \times v^2 + \frac{5 \times 0.98^2}{2 \times 1} = 0.5 \times 9.8 \times 1.98$$

$$v^2 + 10 \times 0.98^2 = 2 \times 9.8 \times 1.98$$

$$v^2 = 2 \times 9.8 \times 1.98 - 10 \times 0.98^2$$

$$v = 5.40$$

The speed of the particle as it passes through its equilibrium position is $5.40\,\text{m}\,\text{s}^{-1}$.

Example 14

A particle P of mass 0.8 kg is attached to one end of a light elastic string of natural length 1 m and modulus 20 N. The other end of the string is fixed to a point A on a rough plane inclined at 30° to the horizontal. P is held at rest at A and then released. P comes to instantaneous rest at a point B 1.5 m from A. Calculate the coefficient of friction between P and the plane.

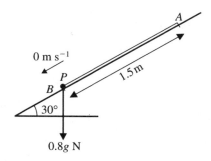

While the particle is moving down the plane there is a frictional force acting up the plane, opposing the motion.

Let this frictional force be F N and the normal reaction between P and the plane be R N.

There will be a tension in the string for part of the motion but this will not affect the calculation of the magnitude of the frictional force.

Resolving perpendicular to the plane:

$$R = 0.8g \cos 30°$$

Using:
$$F = \mu R$$

where μ is the coefficient of friction gives:

$$F = \mu \times 0.8g \cos 30°$$

As the particle travels 1.5 m down the plane, using:

$$\text{work done} = \text{force} \times \text{distance travelled}$$

gives:

work done against frictional force $= \mu \times 0.8 \times 9.8 \cos 30° \times 1.5$ J

The particle starts from rest at A and is at rest again at B. Hence:

$$\text{initial kinetic energy} = \text{final kinetic energy} = 0$$

As the particle moves down the slope from A to B there is a loss of gravitational potential energy. This is given by:

gravitational potential energy lost $= mgh$
$$= 0.8 \times 9.8 \times 1.5 \sin 30° \text{ J}$$

At B, the extension of the string is $(1.5 - 1)$ m $= 0.5$ m.

Using:
$$\text{E.P.E.} = \frac{\lambda x^2}{2l}$$

with $\lambda = 20$ N, $x = 0.5$ m, $l = 1$ m gives:

$$\text{E.P.E.} = \frac{20 \times 0.5^2}{2 \times 1} = 2.5 \text{ J}$$

By the work–energy principle:

work done against frictional force $=$ loss of energy of particle

$$= \text{loss of P.E.} - \text{gain of E.P.E.}$$

Hence:

$$\mu \times 0.8 \times 9.8 \cos 30° \times 1.5 = 0.8 \times 9.8 \times 1.5 \sin 30° - 2.5$$

$$\mu = \frac{0.8 \times 9.8 \times 1.5 \sin 30° - 2.5}{0.8 \times 9.8 \cos 30° \times 1.5}$$

$$\mu = 0.3318$$

So the coefficient of friction between P and the plane is 0.332.

Exercise 2C

Whenever a numerical value of g is required, take $g = 9.8\,\text{m}\,\text{s}^{-2}$.

1 An elastic spring of natural length 1 m and modulus 60 N rests on a smooth horizontal floor with one end fixed. A particle of mass 2 kg is attached to the free end of the spring which is compressed to a length of 0.75 m. The particle is released. Calculate the speed of the particle when the spring returns to its natural length.

2 A particle P of mass 3 kg is attached to one end of a light elastic string of natural length 1.5 m and modulus 6 N. The other end of the string is fixed to a point O on a smooth horizontal table. P is released from rest at a point on the table 3 m from O. Calculate the speed of P (a) when the string becomes slack (b) when $OP = 2$ m.

3 A particle of mass 0.5 kg is attached to one end A of a light elastic string of natural length 2 m and modulus 20 N. The other end B of the string is fixed to a point on the ceiling. The particle is held at a distance of 1.5 m vertically below B and then released. Calculate (a) the length of the string when the particle reaches its lowest point (b) the speed of the particle when it passes through its equilibrium position.

4 A jack-in-the-box is made using a spring of natural length 0.2 m and modulus 100 N and a 'jack' of mass 0.5 kg. When the lid is closed the spring is compressed to a length of 0.1 m. Assuming the spring to be vertical throughout, calculate the maximum distance that the 'jack' will rise when the lid is suddenly raised.

5 A particle P of mass m is attached to one end of a light elastic string of natural length l whose other end is attached to a point A on a ceiling. When P hangs in equilibrium AP has length $\frac{5l}{3}$. Show that if P is projected vertically downwards from A with speed $\sqrt{\left(\frac{3gl}{2}\right)}$, P will first come to instantaneous rest after moving a distance $\frac{10l}{3}$.

6 A particle of mass m is attached to one end of an elastic spring of natural length l and modulus λ. The particle and spring rest on a rough horizontal surface and the other end of the spring is fixed. The coefficient of friction between the particle and the surface is μ. The particle is held at rest with the spring compressed to a length $\frac{2}{3}l$ and then released. Show that the particle is instantaneously at rest when it has moved a distance x, where

$$x = \frac{2\lambda l - 6\mu lmg}{3\lambda}.$$

7 A particle P of mass m is attached to one end of a light elastic string of natural length l and modulus λ. The other end of the string is fixed to a point A on a smooth plane inclined at $\arcsin\frac{3}{5}$ to the horizontal. P is held at rest on the inclined plane with AP along a line of greatest slope of the plane. P is released and first comes to instantaneous rest after moving a distance x.
(a) When P is released from a point below A such that $AP = l$ show that $x = \frac{6mgl}{5\lambda}$.
(b) Given that $\lambda = \frac{6mg}{5}$, find x in terms of l when P is released from a point above A such that $AP = l$.

8 A particle P of mass $2\,\text{kg}$ is attached to one end of a light elastic string of natural length $0.6\,\text{m}$ and modulus $20\,\text{N}$. One end of the string is fixed to a point A on a rough plane inclined at $45°$ to the horizontal. P is held at rest on the plane below A with AP along a line of greatest slope of the plane and $AP = 0.6\,\text{m}$. The coefficient of friction between P and the plane is 0.2. P is released. Calculate the distance P moves down the plane before coming to instantaneous rest. Calculate also the speed of P when $AP = 0.8\,\text{m}$.

9   A particle P of mass 0.3 kg is suspended by two identical elastic strings of natural length 1.5 m and modulus λ N. The other ends of the strings are fixed to two points A and B on a horizontal ceiling where $AB = 3$ m. P is released from rest at the mid-point of AB and falls vertically until it is instantaneously at rest at a point 1 m below the level of the ceiling. Calculate the value of λ and the speed of P when it is 0.5 m below the level of the ceiling.

10 A boy makes a catapult by attaching the ends of a piece of elastic of natural length 0.3 m and modulus 60 N to the ends of a forked stick which are 0.2 m apart. He takes a stone of mass 0.1 kg and places it at the mid-point of the elastic, before pulling the stone backwards until the elastic measures 0.75 m, keeping the elastic horizontal. The stone is then released.
(a) Briefly describe the mathematical model you would use to investigate the motion of the stone, stating any assumptions you would make about the stone or the elastic.
(b) Calculate the speed of the stone on release.
(c) The boy keeps the elastic horizontal. Would the answer to (b) be different if the elastic were vertical? Briefly justify your answer.

11 A light elastic string has natural length l and modulus λ. One end is fixed to a point A on a ceiling and a particle P of mass m is attached to the other end. P is held vertically below A so that $AP = 2l$ and then released. P has speed v when the extension of the string is x. Show that, while the string remains taut,

$$\frac{1}{2}mv^2 = \frac{1}{2}l(\lambda - 2mg) + mgx - \frac{1}{2}\lambda\frac{x^2}{l}$$

By considering the speed of P when $x = 0$ show that the string will never become slack provided $2mg > \lambda$.

Exercise 2D   Mixed questions

Whenever a numerical value of g is required take $g = 9.8\,\mathrm{m\,s^{-2}}$.

1   A light elastic string of modulus 49 N and natural length 1.5 m has one end attached to point A of a ceiling. A particle P of mass 0.5 kg is attached to the other end of the string.

(a) Find the length of the string when P is hanging in equilibrium.

The particle P is released from rest at A and falls vertically. Find

(b) the speed of P as P passes through the equilibrium position

(c) the length of the string when P is momentarily at rest.

2 A light elastic string of natural length 50 cm is stretched between two points A and B on a smooth horizontal table. The distance AB is 75 cm and the tension in the string is 15 N.

(a) Find, in newtons, the modulus of elasticity of the string.

The mid-point of the string is pulled 10 cm towards B. The string still lies along the straight line AB.

(b) Find, in joules, the change in the elastic energy of the string.

(c) State whether the change is an increase or a decrease.

3 The ends of a light elastic string of natural length 0.8 m and modulus 19.6 N are attached to points A and B of a horizontal ceiling where $AB = 1.6$ m. A particle P is attached to the mid-point of the string and hangs in equilibrium 0.6 m below the mid-point of AB. Find

(a) the elastic energy stored in the string

(b) the mass of the particle.

4 A particle P of mass m is attached to one end of a light elastic string of natural length 0.5 m. The other end of the string is attached to a fixed point on a ceiling. When P hangs in equilibrium the length of the string is 0.7 m.

(a) Find, in terms of m and g, the modulus of elasticity of the string.

P is now held in equilibrium by a horizontal force of magnitude $3mg$. Find

(b) the length of the string

(c) the angle, to the nearest degree, between the string and the horizontal.

5 A small ball of mass m is attached to one end of a light elastic string of natural length l and modulus $5mg$. The other end of the string is attached to the base of a vertical wall. The ball is released from rest on a rough horizontal surface at a distance $\frac{3}{2}l$ from the base of the wall. The coefficient of friction between the ball and the surface is $\frac{1}{5}$.

(a) Find, in terms of l and g, the speed of the ball when it hits the wall.

Given that the coefficient of restitution between the ball and the wall is $\frac{2}{5}$, find, in terms of l and g,

(b) the speed of the ball immediately after the impact with the wall

(c) the distance from the wall at which the ball comes to rest.

6 A particle P of mass M is attached to one end of a light elastic string of natural length l and modulus $2Mg$. The other end of the string is attached to a fixed point A of a rough horizontal table. Initially P is at rest at point B of the table where $AB = \frac{4}{3}l$. Given that P is in limiting equilibrium,

(a) find the coefficient of friction between P and the table.

P is now moved away from A and is released from rest at point C of the table where $AC = 2l$. Find:

(b) the speed of P when the string returns to its natural length

(c) the total distance moved by P when P comes to rest.

7 Using Hooke's law, show by integration that the work done in stretching a light elastic string of natural length l and modulus of elasticity λ, from length l to $(l+e)$ is $\dfrac{\lambda e^2}{2l}$.

A light elastic string of natural length 4 metres and modulus of elasticity $4mg$ has one end attached to a fixed point A.
A particle P, of mass m, is attached to the other end of the string.

(a) Show that, when P hangs in equilibrium, the length of AP is $5\,$m.

The particle is released from rest at A and falls vertically.

(b) Find the greatest length of the string.

SUMMARY OF KEY POINTS

1 Elastic strings and springs have a tension or thrust, T, given by:

$$T = \frac{\lambda x}{l}$$

where λ is the modulus of elasticity, x the extension or compression and l the natural length of the string or spring. This is known as Hooke's law.

2 The work done in stretching (or compressing) an elastic string (or spring) with modulus λ from its natural length l to a length $(l + x)$ (or $(l - x)$) is $\frac{\lambda x^2}{2l}$.

3 The elastic energy (E.P.E.) in a string (or spring) of modulus λ extended (or compressed) by a length x beyond its natural length is also $\frac{\lambda x^2}{2l}$.

4 The total change of the mechanical energies (that is kinetic, gravitational potential and elastic potential energies) of a system is equal to the work done by any external forces acting on the system.

Further dynamics

Dynamics is the study of the relationship between the resultant force acting on a particle and the motion of that particle. The study of dynamics was begun in Book M1 and extended in Book M2 but was restricted to constant forces. This work will now be extended to include some cases of variable forces.

3.1 Newton's laws for a particle moving in a straight line when the applied force is variable

Chapter 1 looked at the kinematics of a particle moving in a straight line when the acceleration was changing during the motion. Consideration of Newton's second law, $\mathbf{F} = m\mathbf{a}$, will show you that for a particle of constant mass, the acceleration can only be variable if the resultant force acting on the particle is variable.

Consider a particle of mass m moving under the influence of a force F which is a function of time.

That is:
$$F = \mathrm{F}(t)$$

Using Newton's second law, $F = ma$, gives:
$$ma = \mathrm{F}(t)$$

As:
$$a = \frac{\mathrm{d}v}{\mathrm{d}t}$$

it follows that:
$$m\frac{\mathrm{d}v}{\mathrm{d}t} = \mathrm{F}(t)$$

and so, integrating with respect to time gives:
$$mv = \int \mathrm{F}(t)\mathrm{d}t + c$$

where c is an arbitrary constant.

Similarly, if F is a function of the displacement, that is:
$$\mathrm{F} = \mathrm{G}(x)$$

then using Newton's second law gives:
$$ma = \mathrm{G}(x)$$

In chapter 1 you were shown that

$$a = \frac{\mathrm{d}}{\mathrm{d}x}\left(\frac{1}{2}v^2\right)$$

and so:
$$m\frac{\mathrm{d}}{\mathrm{d}x}\left(\frac{1}{2}v^2\right) = G(x)$$

Integrating with respect to x gives:

$$\frac{1}{2}mv^2 = \int G(x)\mathrm{d}x + k$$

where k is an arbitrary constant.

Example 1

A particle P of mass $2\,\mathrm{kg}$ is moving along a straight line. P is initially at rest at a point O on the line.
The force acting on P has magnitude $(3t^2 + e^t)\,\mathrm{N}$ at time $t\,\mathrm{s}$ and acts in the direction OP. Calculate:

(a) the speed of P when $t = 2$
(b) the distance OP when $t = 4$.

(a) Using Newton's second law, $F = ma$, with $F = (3t^2 + e^t)\,\mathrm{N}$ and $m = 2\,\mathrm{kg}$ gives:

$$3t^2 + e^t = 2a$$

Using:
$$a = \frac{\mathrm{d}v}{\mathrm{d}t}$$

gives:
$$2\frac{\mathrm{d}v}{\mathrm{d}t} = 3t^2 + e^t$$

Integrating with respect to t gives:

$$2v = \frac{3}{3}t^3 + e^t + c$$
$$= t^3 + e^t + c$$

As $v = 0$ when $t = 0$:

$$0 = 0 + 1 + c$$

So:
$$c = -1$$

and:
$$2v = t^3 + e^t - 1 \tag{1}$$

Substituting $t = 2$ gives:
$$2v = 8 + e^2 - 1$$
$$v = \tfrac{7}{2} + \tfrac{1}{2}e^2$$

The speed of P when $t = 2$ is $(3\tfrac{1}{2} + \tfrac{1}{2}e^2)\,\mathrm{m\,s}^{-1}$.

(b) From equation (1) above, $2v = t^3 + e^t - 1$.

Using $v = \dfrac{\mathrm{d}x}{\mathrm{d}t}$ gives:
$$2\frac{\mathrm{d}x}{\mathrm{d}t} = t^3 + e^t - 1$$

Integrating with respect to t gives:

$$2x = \frac{1}{4} t^4 + e^t - t + K$$

As $x = 0$ when $t = 0$:

$$0 = e^0 + K$$

So:

$$K = -1$$

and:

$$2x = \frac{1}{4} t^4 + e^t - t - 1$$

Substituting $t = 4$ gives:

$$2x = \frac{1}{4} \times 4^4 + e^4 - 4 - 1$$

$$2x = 59 + e^4$$

$$x = \frac{59}{2} + \frac{1}{2} e^4$$

The distance OP when $t = 4$ is $(29\frac{1}{2} + \frac{1}{2} e^4)$ m.

Example 2

A particle P of mass 0.5 kg is moving in a straight line. When P is at a distance x m from a fixed point O on the line, the force acting on it is of magnitude $(6 - \cos 3x)$ N and acts in the direction OP. Given that P passes through O with speed $5\,\text{m s}^{-1}$, calculate the speed of P when $x = 2$.

Using Newton's second law, $F = ma$, with $F = (6 - \cos 3x)$ N and $m = 0.5$ kg gives:

$$6 - \cos 3x = 0.5a$$

As the force is a function of x, use:

$$a = \frac{\mathrm{d}}{\mathrm{d}x} \left(\frac{1}{2} v^2 \right)$$

to obtain:

$$0.5 \frac{\mathrm{d}}{\mathrm{d}x} \left(\frac{1}{2} v^2 \right) = 6 - \cos 3x$$

Integrating with respect to x gives:

$$0.5 \times \frac{1}{2} v^2 = 6x - \frac{1}{3} \sin 3x + c$$

Since $v = 5$ when $x = 0$:

$$\frac{1}{2} \times \frac{1}{2} \times 5^2 = 0 - 0 + c$$

So:

$$c = \frac{25}{4}$$

And so:

$$\frac{1}{4} v^2 = 6x - \frac{1}{3} \sin 3x + \frac{25}{4}$$

Substituting $x = 2$ gives:

$$\frac{1}{4} v^2 = 6 \times 2 - \frac{1}{3} \sin (3 \times 2) + \frac{25}{4}$$

$$v = 8.565$$

The speed of P when $x = 2$ is $8.57\,\mathrm{m\,s^{-1}}$.

Impulse and momentum

The impulse–momentum equation $\mathbf{F}t = m\mathbf{v} - m\mathbf{u}$ was derived in Book M1 chapter 5 for a constant force $\mathbf{F}$ acting on a particle of constant mass m. For a variable force, $F(t)$, Newton's second law, $F(t) = ma$, can be written:

$$F(t) = m \frac{dv}{dt}$$

Suppose the particle has constant mass and a velocity U at time t_1 and a velocity V at time t_2. Then integrating with respect to t over the interval from t_1 to t_2 gives:

$$\int_{t_1}^{t_2} F(t)dt = m \int_{U}^{V} 1\, dv$$

$$= m[v]_U^V$$

$$= m(V - U)$$

$$= mV - mU$$

You should recognise $mV - mU$ as the change of momentum of the particle. The quantity $\int_{t_1}^{t_2} F(t)\, dt$ is the impulse of the variable force on the particle. Thus the equation:

$$\text{impulse} = \text{change of momentum}$$

is still valid but for a variable force the impulse is given by:

$$\text{impulse} = \int_{t_1}^{t_2} F\, dt$$

Example 3

A stone of mass $0.25\,\mathrm{kg}$ is free to move on a smooth horizontal floor. At time t seconds, $t \geqslant 0$, a horizontal force of magnitude $3t^2\,\mathrm{N}$ and constant direction acts on the stone. When $t = 0$ the stone has speed $8\,\mathrm{m\,s^{-1}}$ in the same direction as the force. When $t = T$ the speed of the stone is $40\,\mathrm{m\,s^{-1}}$.

(a) Choose suitable models for the stone and the floor so that you can find the value of T.
(b) Calculate the value of T.

(a) The stone can be assumed to be small and hence modelled as a particle.

The floor is said to be smooth and horizontal and so can be modelled as a smooth horizontal surface.

(b) Using $\int_{t_1}^{t_2} F \, dt = mV - mU$

with $F = 3t^2$ N, $V = 40$ m s^{-1}, $U = 8$ m s^{-1}, $t_1 = 0$ s, $t_2 = T$ s and $m = 0.25$ kg gives:

$$\int_0^T 3t^2 \, dt = 0.25 \times 40 - 0.25 \times 8$$

$$[t^3]_0^T = 10 - 2$$

$$T^3 = 8$$

$$T = 2$$

Work and energy

When a *constant* force **F** acts on a particle of mass m and moves it through a distance s in the direction of the force, the work done by the force is defined by:

$$Work \ done = F \times s$$

In chapter 3 of Book M2 you saw that:

$$Fs = \tfrac{1}{2} mv^2 - \tfrac{1}{2} mu^2$$

where u and v are the initial and final speeds respectively of the particle.

When the force F acting on the particle is *variable* and a function of x, using Newton's second law with $a = v \dfrac{dv}{dx}$ (see chapter 1) gives:

$$F(x) = mv \frac{dv}{dx} \qquad (1)$$

If the action of the force causes the particle to increase its speed from U to V while moving from the point where $x = x_1$, to the point where $x = x_2$, then integrating equation (1) with respect to x gives:

$$\int_{x_1}^{x_2} F(x) dx = m \int_U^V v \, dv = m \left[\tfrac{1}{2} v^2 \right]_U^V$$

$$\int_{x_1}^{x_2} F(x) dx = \tfrac{1}{2} mV^2 - \tfrac{1}{2} mU^2 \qquad (2)$$

The right hand side of equation (2) is the increase in the kinetic energy of the particle and the left hand side is the work done by the variable force. Thus the equation:

$$work \ done \ by \ force = increase \ in \ kinetic \ energy$$

is still valid but for a variable force $F(x)$ the work done by the force is given by:

■ $$\textbf{work done} = \int_{x_1}^{x_2} \textbf{F}(x)\textbf{d}x$$

Example 4

A particle P of mass $0.5\,\text{kg}$ is moving along the x-axis. When P is a distance $x\,\text{m}$ from the origin, O, a force of magnitude $(10 + 5\text{e}^{-x})\,\text{N}$ acts on P in the direction of the positive x-axis. Calculate the work done by the force in moving P from O to the point A where $x = 2$. Given that P passes through the origin with speed $3\,\text{m s}^{-1}$ calculate the speed of P as it passes through A.

Using: $$\text{work done} = \int_{x_1}^{x_2} F\,\text{d}x$$

with $x_1 = 0$, $x_2 = 2$ and $F = (10 + 5\text{e}^{-x})$ gives:

$$\begin{aligned}
\text{work done} &= \int_0^2 (10 + 5\text{e}^{-x})\text{d}x \\
&= [10x - 5\text{e}^{-x}]_0^2 \\
&= (20 - 5\text{e}^{-2}) - (0 - 5\text{e}^0) \\
&= 20 - 5\text{e}^{-2} + 5 \\
&= 25 - 5\text{e}^{-2} \\
&= 24.32
\end{aligned}$$

The work done is $24.3\,\text{J}$.

Let the speed of P as it passes through A be $v\,\text{m s}^{-1}$.

Using: $$\text{work done} = \text{increase in kinetic energy}$$
$$= \tfrac{1}{2}mv^2 - \tfrac{1}{2}mu^2$$

with   work done $= 24.32\,\text{J}$, $m = 0.5\,\text{kg}$, $u = 3\,\text{m s}^{-1}$ gives:

$$24.32 = \tfrac{1}{2} \times 0.5v^2 - \tfrac{1}{2} \times 0.5 \times 9$$
$$\tfrac{1}{2} \times 0.5v^2 = 24.32 + \tfrac{1}{2} \times 0.5 \times 9$$
$$v^2 = \frac{24.32 + \tfrac{1}{2} \times 0.5 \times 9}{\tfrac{1}{2} \times 0.5}$$
$$v = 10.30$$

The speed of P as it passes through A is $10.3\,\text{m s}^{-1}$.

Exercise 3A

1 A particle P of mass $0.5\,\text{kg}$ is moving in a straight line. P is initially at rest at a point O of the line. The force acting on P has magnitude $4\sin t\,\text{N}$ at time $t\,\text{s}$ and acts in the direction OP.

Calculate (a) the speed of P when $t = 1$ (b) the speed of P when $t = 3$ (c) the time when P first comes to instantaneous rest (d) the distance OP when $t = 3$ (e) the distance OP when P first comes to rest.

2 A particle P of mass 1.5 kg is moving in a straight line. At time t s after it passes through a fixed point O in that line the distance OP is x m and the force acting on P has magnitude $(3 + 3e^{-2x})$ N and acts in the direction OP. Given that P passes through O with speed 2 m s^{-1} calculate the speed of P when (a) $x = 2$, (b) $x = 5$.

3 A particle P of mass 0.5 kg moves along the x-axis under the action of a force whose magnitude at time t s after P passes through the origin is given by $(4t^3 + 3t^2)$ N and which acts in the direction OP. Given that P passes through the origin with velocity 6 m s^{-1} calculate (a) the velocity of P when $t = 2$ (b) the displacement of P when $t = 1$.

4 A particle P of mass 2 kg moves along the x-axis under the action of a force whose magnitude at time t s after P passes through the origin is $(4 + 12t)$ N and which acts in the direction PO. P passes through O with a velocity 20 m s^{-1}. Calculate (a) the speed of P when $t = 2$ (b) the displacement of P when $t = 2$ (c) the time when P has velocity $\frac{25}{3}$ m s^{-1} (d) the displacement of P when P's velocity is $\frac{25}{3}$ m s^{-1}.

5 A particle P of mass 1.5 kg moves in a straight line through a fixed point O. At time t s after passing through O the distance of P from O is x m and the force acting on P has magnitude $(3x + 6)$ N and is directed away from O. Given that P passes through O with speed $2\sqrt{2}$ m s^{-1} calculate (a) the speed of P when $x = 5$ (b) the value of t when $x = 20$.

6 Calculate the magnitude of the impulse of a force of magnitude F N acting from time t_1 seconds to time t_2 seconds where

(a) $F = t^3 + 2t$ $t_1 = 1$ $t_2 = 3$

(b) $F = \dfrac{1}{2t + 4}$ $t_1 = 0$ $t_2 = 4$

(c) $F = 5 + e^{-2t}$ $t_1 = 0$ $t_2 = 3$

(d) $F = 3\cos 5t$ $t_1 = 0$ $t_2 = \dfrac{\pi}{10}$

7 Calculate the work done by a force of magnitude F N directed along the x-axis which moves a particle from $x = x_1$ m to $x = x_2$ m where

(a) $F = x^2 + 2x$ $x_1 = 0$ $x_2 = 5$

(b) $F = 2 + e^{-x}$ $x_1 = 2$ $x_2 = 6$

(c) $F = \dfrac{2}{x} + 3$ $x_1 = 2$ $x_2 = 4$

8 A particle P of mass 2 kg is moving in a straight line. P is initially at rest at a point O of the line. The force acting on P has magnitude $(6t + 4)$ N, where t s is the time for which the force has been acting, and is directed along OP. Calculate (a) the impulse exerted by the force between the times $t = 1$ and $t = 3$ (b) the speed of P when $t = 4$. Given that the final speed of the particle is 66 m s^{-1} calculate (c) the time for which the force acts.

9 A particle P of mass 0.8 kg moves in a straight line through a fixed point O. At time t s after passing through O the distance of P from O is x m and the acceleration of P is $\frac{1}{3}(x^2 + x)$ m s^{-2}. Write down, in terms of x, an expression for the force acting on P and calculate the work done by that force in moving P from $x = 0$ to $x = 5$.

3.2 Newton's law of gravitation

In Newton's day most scientists studied astronomy. Newton's study of planetary motion led him to propose the following **law of gravitation**:

■ **The force of attraction between two bodies of masses M_1 and M_2 is directly proportional to the product of their masses and inversely proportional to the square of the distance between them.**

This statement can be expressed mathematically by the equation

$$F = \frac{GM_1 M_2}{d^2}$$

where G is a constant known as the **constant of gravitation**.

It is because of this force that bodies fall to the earth and the moon orbits the earth.

The numerical value of G was first determined approximately by Henry Cavendish in 1798. Scientists have since improved Cavendish's approximation and the numerical approximation used today, when the force is measured in newtons, the distance in metres and the masses in kilograms, is 6.67×10^{-11}.

Units of G

From the equation:
$$F = G \frac{M_1 M_2}{d^2}$$

it follows that:
$$G = \frac{Fd^2}{M_1 M_2} = Fd^2 (M_1 M_2)^{-1} \tag{1}$$

With the usual S.I. units F is in newtons, d is in metres and M_1 and M_2 are in kilograms.

Using Newton's second law, $\mathbf{F} = m\mathbf{a}$, and the fact that the units on both sides of an equation must be the same, it follows that 1 newton must be $1 \, \text{kg m s}^{-2}$.

The units on both sides of equation (1) above must also be the same so the units for G must be:

$$\text{kg m s}^{-2} \, \text{m}^2 \, \text{kg}^{-2}$$

which simplifies to:

$$\text{kg}^{-1} \, \text{m}^3 \, \text{s}^{-2}$$

Hence G is $6.67 \times 10^{-11} \, \text{kg}^{-1} \, \text{m}^3 \, \text{s}^{-2}$

Example 5

Two particles of masses 2 kg and 3 kg are 10 cm apart. Determine the magnitude of the gravitational force between them.

Using:
$$F = G \frac{M_1 M_2}{d^2}$$

with $G = 6.67 \times 10^{-11} \, \text{kg}^{-1} \, \text{m}^3 \, \text{s}^{-2}$, $M_1 = 2 \, \text{kg}$, $M_2 = 3 \, \text{kg}$ and $d = 10 \, \text{cm} = 0.1 \, \text{m}$ gives:

$$F = \frac{6.67 \times 10^{-11} \times 2 \times 3}{0.1^2}$$
$$F = 4.002 \times 10^{-8}$$

The magnitude of the gravitational force between the particles is $4.0 \times 10^{-8} \, \text{N}$.

It can be seen from the above example that the gravitational attraction between everyday objects is very small compared to the other forces acting and so this force may be ignored in calculations.

Relationship between *G* and *g*

Consider a particle of mass m which is at rest on the surface of the earth. Newton proved that the gravitational attraction due to the earth is the same as that of a particle whose mass is equal to the earth's mass and is situated at the centre of the earth. So a particle model may be used for the earth.

Suppose the mass of the earth is M_e kg and its radius is R m. By Newton's law of gravitation the magnitude of the force of attraction between the particle and the earth is:

$$F = \frac{GmM_e}{R^2} \tag{1}$$

However, as you saw in Book M1, the force with which the earth attracts the particle is the weight of the particle, so:

$$F = mg \tag{2}$$

Combining equations (1) and (2) gives:

$$mg = \frac{GmM_e}{R^2}$$

or:

$$g = \frac{GM_e}{R^2}$$

As the force of attraction between two bodies of masses M_1 and M_2 is given by

$$F = \frac{GM_1M_2}{d^2}$$

it follows that if the distance d between the two bodies is changing, the force is a function of d.

Example 6

A rocket of mass M is launched from a point on the surface of the earth and travels vertically upwards. The engine section is detached when the rocket is at a distance d from the centre of the earth and travelling at speed u. When the rocket is at a distance x, $x > d$, from the centre of the earth the speed of the rocket is v and the gravitational attraction of the earth on the rocket is $\dfrac{KM}{x^2}$

where K is a positive constant. It may be assumed that, after the engine section is detached, all forces on the rocket other than the gravitational attraction of the earth can be ignored. Find an expression for v.

The force acting on the rocket has magnitude $\dfrac{KM}{x^2}$ and is directed towards the earth. The positive direction is that of increasing x, that is away from the earth. So, when writing down the equation of motion, a negative sign is required with the force.

Using:
$$F = Ma$$

gives:
$$-\frac{KM}{x^2} = Ma$$

As the force is a function of x, use $a = v\dfrac{\mathrm{d}v}{\mathrm{d}x}$ to obtain
$$-\frac{K}{x^2} = v\frac{\mathrm{d}v}{\mathrm{d}x}$$

Separating the variables gives:
$$-K\int \frac{\mathrm{d}x}{x^2} = \int v\,\mathrm{d}v$$
$$-K\int x^{-2}\,\mathrm{d}x = \int v\,\mathrm{d}v$$
$$Kx^{-1} = \tfrac{1}{2}v^2 + c$$
$$\frac{K}{x} = \tfrac{1}{2}v^2 + c$$

When $x = d$, $v = u$.

So:
$$\frac{K}{d} = \tfrac{1}{2}u^2 + c$$
$$c = \frac{K}{d} - \tfrac{1}{2}u^2$$

Hence:
$$\frac{K}{x} = \tfrac{1}{2}v^2 + \frac{K}{d} - \tfrac{1}{2}u^2$$
$$v^2 = \frac{2K}{x} - \frac{2K}{d} + u^2$$
$$v = \sqrt{\frac{2K}{x} - \frac{2K}{d} + u^2}$$

Example 7

A rocket of mass 8000 kg is fired vertically upwards with speed $u\,\mathrm{m\,s}^{-1}$ from a point on the surface of the earth. By modelling the earth as sphere of radius 6.37×10^6 m and mass 5.96×10^{24} kg and assuming that all forces on the rocket apart from the force of attraction from the earth are negligible, calculate:

(a) the minimum value of u if the rocket is to reach a height of 8×10^6 m above the surface of the earth
(b) the escape velocity of the rocket.

By Newton's law of gravitation, the magnitude of the force on the rocket when it is at a distance x from the centre of the earth is

$$\frac{G \times 8000 \times 5.96 \times 10^{24}}{x^2}$$

and the force is directed towards the earth.

Using: $$F = ma$$

gives: $$-\frac{G \times 8000 \times 5.96 \times 10^{24}}{x^2} = 8000a$$

As the acceleration is a function of the distance, use $a = v\dfrac{\mathrm{d}v}{\mathrm{d}x}$.

Hence: $$-\frac{G \times 5.96 \times 10^{24}}{x^2} = v\frac{\mathrm{d}v}{\mathrm{d}x}$$

Separating the variables gives:

$$-G \times 5.96 \times 10^{24} \int \frac{1}{x^2}\,\mathrm{d}x = \int v\,\mathrm{d}v$$

$$-G \times 5.96 \times 10^{24} \int x^{-2}\,\mathrm{d}x = \tfrac{1}{2}v^2 + c$$

Hence: $$G \times 5.96 \times 10^{24} x^{-1} = \tfrac{1}{2}v^2 + c$$

The rocket is fired with speed u from the surface of the earth, so $x = 6.37 \times 10^6$ when $v = u$.

So $$\frac{G \times 5.96 \times 10^{24}}{6.37 \times 10^6} = \tfrac{1}{2}u^2 + c$$

and: $$\tfrac{1}{2}v^2 = \frac{G \times 5.96 \times 10^{24}}{x} - \frac{G \times 5.96 \times 10^{24}}{6.37 \times 10^6} + \tfrac{1}{2}u^2 \qquad (1)$$

As $\dfrac{G \times 800 \times 5.96 \times 10^{24}}{x}$ is a decreasing function of x, the speed is decreasing as x increases.

(a) For the minimum value of u to reach any given height, the speed at that height will be zero. So, using $v = 0$ when $x = (8 \times 10^6 + 6.37 \times 10^6) = 1.437 \times 10^7$ will enable the required minimum value of u to be calculated.

Thus:
$$0 = \frac{G \times 5.96 \times 10^{24}}{1.437 \times 10^7} - \frac{G \times 5.96 \times 10^{24}}{6.37 \times 10^6} + \tfrac{1}{2} u_{min}^2$$

As $G = 6.67 \times 10^{-11} \, \text{kg}^{-1} \, \text{m}^3 \, \text{s}^{-2}$, this gives:

$$\tfrac{1}{2} u_{min}^2 = 6.67 \times 10^{-11} \left(\frac{5.96 \times 10^{18}}{6.37} - \frac{5.96 \times 10^{17}}{1.437} \right)$$

$$u_{min}^2 = 2 \times 6.67 \left(\frac{5.96 \times 10^7}{6.37} - \frac{5.96 \times 10^6}{1.437} \right)$$

$$u_{min} = 8.34 \times 10^3$$

The minimum speed required is $8.34 \times 10^3 \, \text{m s}^{-1}$.

(b) The escape velocity is the minimum velocity at which the rocket must be fired if its subsequent velocity is never to reach zero.

That is, $v^2 > 0$ for all values of x.

Using equation (1):

$$\tfrac{1}{2} u^2 + \frac{G \times 5.96 \times 10^{24}}{x} > \frac{G \times 5.96 \times 10^{24}}{6.37 \times 10^6} \qquad \text{for all values of } x.$$

The limiting value of $\dfrac{1}{x}$ as x becomes very large is 0.

So u satisfies:
$$u^2 > 2 \times \frac{G \times 5.96 \times 10^{24}}{6.37 \times 10^6}$$

That is:
$$u^2 > 2 \times \frac{6.67 \times 10^{-11} \times 5.96 \times 10^{24}}{6.37 \times 10^6}$$

$$u > 1.12 \times 10^4$$

Hence the escape velocity is $1.12 \times 10^4 \, \text{m s}^{-1}$.

Exercise 3B

Whenever a numerical value of g or G is required take $g = 9.8 \, \text{m s}^{-2}$ and $G = 6.67 \times 10^{-11} \, \text{kg}^{-1} \, \text{m}^3 \, \text{s}^{-2}$.

1 Two particles have masses 0.2 kg and 0.5 kg. Calculate the magnitude of the gravitational attraction between them when they are (a) 2 cm apart (b) 2 m apart (c) 2 km apart.

2 Two identical particles which are 5 cm apart experience a gravitational attraction of magnitude 7.6×10^{-9} N. Calculate the masses of the particles.

3 The acceleration due to the earth's gravity is taken to be $9.8 \, \mathrm{m \, s^{-2}}$ at sea level. Obtain an estimate for this acceleration at the top of a mountain which is 7 km above sea level. You may assume the earth to be a sphere of radius 6370 km. State clearly any other assumptions you make.

4 Assuming the earth to be a sphere of radius 6370 km, obtain an estimate for the mass of the earth.

5 The moon can be assumed to be a sphere of radius 1740 km and mass 7.34×10^{22} kg. An astronaut of mass 90 kg is making a moon walk. Calculate the magnitude of the force of attraction between the moon and the astronaut.

6 Assuming the moon to be a sphere of radius 1740 km and mass 7.34×10^{22} kg, obtain an estimate for the acceleration due to the moon's gravity. State clearly the model you have used for the moon.

7 The moon can be modelled as a sphere of radius 1740 km and mass 7.34×10^{22} kg. A rocket of mass m is fired vertically from the surface of the moon with speed $1.5 \, \mathrm{km \, s^{-1}}$. It may be assumed that all forces on the rocket other than the gravitational attraction of the moon can be ignored.
(a) Find the speed of the rocket when it is at a height of 10 km above the surface of the moon.
(b) Find the maximum distance from the moon's surface reached by the rocket.

8 The moon can be modelled as a sphere of radius 1740 km and mass 7.34×10^{22} kg. Find the escape velocity of a rocket fired vertically from a point on the moon's surface. You may assume that all forces on the rocket other than the gravitational attraction of the moon are negligible.

3.3 Simple harmonic motion

We now consider a very special and important kind of motion of a particle P moving on a straight line, called **simple harmonic motion** or **S.H.M.** In this motion the acceleration is always *towards* a fixed point O of the line and is proportional to the displacement of P from O. The fixed point O is often called the **centre of oscillation**. To ensure that the acceleration of the particle is always towards O, this acceleration is usually written as:

$$\ddot{x} = -\omega^2 x$$

where ω is a real number.

As $\ddot{x}$ is the acceleration it follows that:

$$\ddot{x} = \frac{dv}{dt}$$

but as shown in chapter 1:

$$\frac{dv}{dt} = v\frac{dv}{dx}$$

So:

$$\ddot{x} = v\frac{dv}{dx} = -\omega^2 x$$

or:

$$\frac{d}{dx}\left(\frac{1}{2}v^2\right) = -\omega^2 x$$

Integrating with respect to x gives:

$$\tfrac{1}{2}v^2 = \int -\omega^2 x\,dx + c$$

where c is an arbitrary constant.

So:

$$\tfrac{1}{2}v^2 = -\omega^2 \times \tfrac{1}{2}x^2 + c$$

$$v^2 = -\omega^2 x^2 + 2c$$

Since the speed of P is the modulus of v, or in other words the modulus of $\frac{dx}{dt}$, it follows that the speed of P is zero when x has its maximum or minimum value. Hence c can be determined by substituting $v = 0$ and $x = a$, where a is the maximum displacement of the particle from O.

Hence:

$$0 = -\omega^2 a^2 + 2c$$

$$2c = \omega^2 a^2$$

and so:

$$v^2 = -\omega^2 x^2 + \omega^2 a^2$$

■

$$v^2 = \omega^2\left(a^2 - x^2\right) \tag{1}$$

This is a standard result which should be memorised. It may be quoted and used in the examination without first proving it.

As $v = \dfrac{dx}{dt}$, an expression for x can be found by integrating equation (1).

From equation (1): $\qquad\qquad \dfrac{dx}{dt} = \pm\omega(a^2 - x^2)^{\frac{1}{2}}$

Separating the variables and integrating gives:

$$\int \frac{dx}{(a^2 - x^2)^{\frac{1}{2}}} = \pm \int \omega \, dt$$

For the left hand integral a substitution is required as shown in Book P3 chapter 4.

Substitute: $\qquad\qquad\qquad\qquad x = a\sin\theta$

So: $\qquad\qquad\qquad\qquad dx = a\cos\theta \, d\theta$

and: $\qquad\qquad (a^2 - x^2)^{\frac{1}{2}} = (a^2 - a^2\sin^2\theta)^{\frac{1}{2}}$

$$= a\cos\theta$$

since $\cos^2\theta + \sin^2\theta = 1$ (Book P1 chapter 2).

Hence: $\qquad\qquad \displaystyle\int \frac{dx}{(a^2 - x^2)^{\frac{1}{2}}} = \int \frac{a\cos\theta \, d\theta}{a\cos\theta}$

$$= \int d\theta$$

So: $\qquad\qquad\qquad \displaystyle\int d\theta = \pm \int \omega \, dt$

$$\theta = \pm\omega t + \alpha$$

and: $\qquad\qquad\qquad \sin\theta = \sin(\omega t + \alpha)$

or: $\qquad\qquad\qquad x = a\sin(\omega t + \alpha) \qquad\qquad (2)$

This function has period $\dfrac{2\pi}{\omega}$ and maximum and minimum values $\pm a$, so $-a \leqslant x \leqslant a$. a is called the **amplitude** of the motion.

To investigate the motion further, different values of the constant of integration, α, must be considered.

Case (i) $\alpha = 0$

When $\alpha = 0$, equation (2) becomes:

$$x = a \sin \omega t$$

The graph of $x = a \sin \omega t$ is:

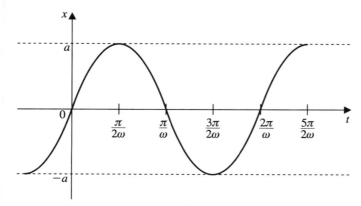

The graph shows that $x = a \sin \omega t$ gives the displacement from the centre of oscillation of a particle moving with S.H.M. of amplitude a and period $\dfrac{2\pi}{\omega}$ which is at the centre of the oscillation (that is $x = 0$) when $t = 0$.

Case (ii) $\alpha = \frac{\pi}{2}$

when $\alpha = \frac{\pi}{2}$, equation (2) becomes

$$x = a \sin\left(\omega t + \frac{\pi}{2}\right)$$

In Book P1 chapter 2 it was shown that the graph of $x = a \sin(\omega t + \frac{\pi}{2})$ is the same shape as that of $x = a \sin \omega t$ but is translated $\frac{\pi}{2}$ to the left.

Hence the graph of $x = a \sin\left(\omega t + \frac{\pi}{2}\right)$ is:

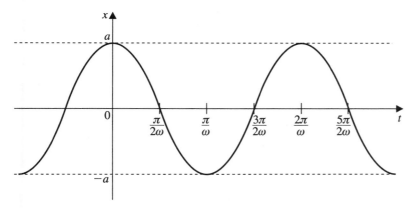

which you will recognise as being the graph of $x = a \cos \omega t$.

Hence the equation $x = a \cos \omega t$ gives the displacement from the centre of oscillation of a particle moving with S.H.M. of amplitude a and period $\dfrac{2\pi}{\omega}$ which has x maximum when $t = 0$.

Case (iii) α is neither of the above.

When α takes some value other than 0 or $\frac{\pi}{2}$ the graph of $x = a \sin(\omega t + \alpha)$ is a translation of the graph of $x = a \sin \omega t$ through a distance α to the left.

The graph of $x = a \sin(\omega t + \alpha)$ is therefore:

> Note: In your M3 exam you will only need to answer questions where $\alpha = 0$ or $\frac{\pi}{2}$.
>
> That is, where
> $$x = a \sin \omega t$$
> or $$x = a \cos \omega t$$

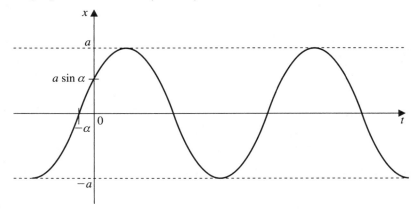

$x = a \sin(\omega t + \alpha)$ gives the solution when the particle is neither at the centre nor at an extreme point of the oscillation when $t = 0$.

Thus for S.H.M. of amplitude a defined by the equation $\ddot{x} = -\omega^2 x$:
$$v^2 = \omega^2(a^2 - x^2)$$
and: $x = a \sin \omega t$ if $x = 0$ when $t = 0$

$x = a \cos \omega t$ if $x = a$ when $t = 0$

$x = a \sin(\omega t + \alpha)$ if x is at some other point when $t = 0$.

Example 8
A particle P is moving along a straight line with S.H.M. of amplitude $0.5\,\text{m}$. It passes through the centre of the oscillation O with speed $5\,\text{m s}^{-1}$. Calculate the period of the oscillation and the speed of the particle when $OP = 0.25\,\text{m}$.

Using: $v^2 = \omega^2(a^2 - x^2)$

with $a = 0.5\,\text{m}$, $v = 5\,\text{m s}^{-1}$ and $x = 0\,\text{m}$ gives:
$$5^2 = \omega^2(0.5^2 - 0)$$
$$\omega^2 = \frac{5^2}{0.5^2}$$
$$\omega = \frac{5}{0.5} = 10$$

Using:

$$\text{period} = \frac{2\pi}{\omega}$$

gives:

$$\text{period} = \frac{2\pi}{10} = \frac{\pi}{5}$$

The period of oscillation is $\frac{\pi}{5}$ seconds.

Using:

$$v^2 = \omega^2(a^2 - x^2)$$

with $a = 0.5\,\text{m}$, $x = 0.25\,\text{m}$ and $\omega = 10$ gives:

$$v^2 = 10^2(0.5^2 - 0.25^2)$$

$$v = 10\sqrt{(0.5^2 - 0.25^2)}$$

$$v = 4.33$$

The speed of P when $OP = 0.25\,\text{m}$ is $4.33\,\text{m}\,\text{s}^{-1}$.

Example 9

A particle is moving in a straight line with S.H.M. Its maximum speed is $5\,\text{m}\,\text{s}^{-1}$ and its maximum acceleration is $12\,\text{m}\,\text{s}^{-2}$. Calculate the amplitude and period of the oscillation.

The maximum speed occurs when the acceleration is zero. Since $\ddot{x} = -\omega^2 x$ it occurs when $x = 0$. The acceleration has its maximum magnitude when x has its maximum magnitude. Hence it occurs when $x = \pm a$.

Using:

$$v^2 = \omega^2(a^2 - x^2)$$

with $v = 5\,\text{m}\,\text{s}^{-1}$ and $x = 0$ gives:

$$5^2 = \omega^2 a^2 \qquad (1)$$

Using:

$$\ddot{x} = -\omega^2 x$$

gives:

$$|\ddot{x}| = |-\omega^2 x|$$

And so:

$$|\ddot{x}| = \omega^2 |x|$$

As the maximum magnitude for $\ddot{x}$ is $12\,\text{m}\,\text{s}^{-2}$ and this occurs when $x = \pm a$, it follows that

$$12 = \omega^2 a \qquad (2)$$

Dividing (1) by (2) gives:

$$\frac{5^2}{12} = \frac{\omega^2 a^2}{\omega^2 a}$$

$$a = \frac{5^2}{12} = 2.083$$

The amplitude is $2.08\,\text{m}$.

Substituting $a = \dfrac{25}{12}$ in equation (2) gives:

$$\frac{25}{12}\,\omega^2 = 12$$

$$\omega^2 = \frac{12^2}{25}$$

$$\omega = \frac{12}{5} = 2.4$$

Hence:
$$\text{period} = \frac{2\pi}{\omega} = \frac{2\pi}{2.4} = 2.617$$

The period is 2.62 s.

There are many real-life situations that involve oscillatory motion. Examples include the motion of the pistons in an engine, the swinging pendulum of a clock and the motion of an object bobbing up and down on the waves. A particle moving with S.H.M. can provide a good model for the mathematical analysis of such motions.

Example 10

A buoy on the surface of the sea is tied to a pier and oscillates in a vertical line with S.H.M. The period of oscillation is $\frac{3\pi}{4}$ seconds and the maximum speed of the buoy is $4\,\mathrm{m\,s}^{-1}$. At the highest point of the motion the buoy is 5 m below the pier. Choose a suitable model for the buoy and hence calculate the amplitude of the motion and find the time taken by the buoy to fall from its highest point to 6 m below the pier.

The buoy will be small compared with other measurements and so can be modelled as a particle.

Using:
$$\text{period} = \frac{2\pi}{\omega}$$

with period $= \dfrac{3\pi}{4}$ gives:
$$\frac{3\pi}{4} = \frac{2\pi}{\omega}$$

$$\omega = \frac{8}{3}$$

Using:
$$v^2 = \omega^2(a^2 - x^2)$$

with $\omega = \frac{8}{3}$, $v = 4$, $x = 0$ gives:

$$4^2 = \frac{8^2}{3^2} \times a^2$$

$$a^2 = \frac{4^2 \times 3^2}{8^2}$$

$$a = \frac{4 \times 3}{8} = 1.5$$

The amplitude is 1.5 m.

Taking $t = 0$ at the highest point, the form required is $x = a\cos\omega t$.

At the highest point, the buoy is 5 m below the pier. To find the time when the buoy is 6 m below the pier you need to find the time when the buoy has fallen a distance $(6 - 5)\,\text{m} = 1\,\text{m}$.

So when the buoy has fallen 1 m, $x = 1.5 - 1 = 0.5\,\text{m}$.

Using:
$$x = a\cos\omega t$$

with $a = 1.5\,\text{m}$, $\omega = \frac{8}{3}$ and $x = 0.5\,\text{m}$ gives:

$$0.5 = 1.5\cos\tfrac{8}{3}t$$
$$\tfrac{1}{3} = \cos\tfrac{8}{3}t$$
$$\tfrac{8}{3}t = \arccos\left(\tfrac{1}{3}\right)$$
$$t = \tfrac{3}{8}\arccos\left(\tfrac{1}{3}\right)$$
$$t = 0.4616$$

The time taken is 0.462 s.

A geometrical approach to simple harmonic motion

Consider a particle P which is moving round a circle of radius a, centre the origin O, with a constant angular speed ω in an anticlockwise sense. Let A be the foot of the perpendicular from P onto the x-axis.

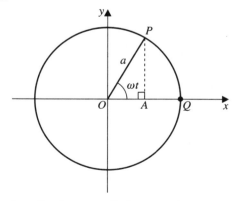

If P is at the point Q when $t = 0$ then at time t seconds:

$$\angle POA = \omega t$$

and so:
$$OA = x = a\cos\omega t$$

Differentiating with respect to time gives:

$$\dot{x} = -a\omega \sin \omega t$$

and:

$$\ddot{x} = -a\omega^2 \cos \omega t$$

So:

$$\ddot{x} = -\omega^2 x$$

The point A is therefore moving along the x-axis with S.H.M. of amplitude a and period $\dfrac{2\pi}{\omega}$.

This connection between circular motion and S.H.M. can be particularly useful when calculating the time for a particle to move between two points of the oscillation. The circle associated with any particular S.H.M. is often called the **reference circle**.

Example 11

A particle P is moving along a straight line with S.H.M. centre O with a period 5 s and amplitude 0.4 m. A and B are points on the line of P's motion such that OA is 0.3 m and OB is 0.2 m with A and B on opposite sides of O. Calculate the time taken by P to move directly from A to B.

First draw a diagram showing a circle of radius 0.4 m with a diameter representing the line on which P is moving. The points A and B are on this diameter on opposite sides of O as shown:

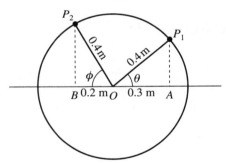

The time taken by P to move directly from A to B is the time taken by a particle moving on the reference circle with angular speed ω to move from P_1 to P_2; that is, the time to move round an arc of the circle which subtends an angle of $(\pi - \theta - \phi)^c$ at O.

From the diagram:
$$\theta = \arccos\left(\frac{0.3}{0.4}\right) = \arccos\left(\frac{3}{4}\right)$$

$$\phi = \arccos\left(\frac{0.2}{0.4}\right) = \arccos\left(\frac{1}{2}\right)$$

So the required angle is:

$$\pi - \theta - \phi = \pi - \arccos\left(\tfrac{3}{4}\right) - \arccos\left(\tfrac{1}{2}\right)$$

The particle on the reference circle takes 5 s (the period) to travel a distance corresponding to an angle of 2π. Let the time taken to travel $\pi - \theta - \phi$ be T s.

Since the angular speed is constant:

$$\frac{T}{\pi - \theta - \phi} = \frac{5}{2\pi}$$

$$T = \frac{5}{2\pi} \times \left(\pi - \arccos\left(\tfrac{3}{4}\right) - \arccos\left(\tfrac{1}{2}\right)\right)$$

$$T = 1.091$$

The time to travel from A to B is 1.09 s.

Exercise 3C

1 A particle P is moving on a straight line with S.H.M. of amplitude 0.3 m. It passes through the centre of the oscillation O with speed $2\,\text{m s}^{-1}$. Calculate the period of oscillation and the speed of the particle when $OP = 0.1$ m.

2 A particle P is moving on a straight line with S.H.M. of amplitude 0.5 m and period $\dfrac{2\pi}{5}$ s. Calculate the maximum speed of the particle.

3 A particle P is moving on a straight line with S.H.M. of period $\dfrac{\pi}{3}$ s. Its maximum speed is $5\,\text{m s}^{-1}$. Calculate the amplitude of the motion and the speed of P 0.2 s after passing through the centre of oscillation.

4 A particle is moving on a straight line with S.H.M. Its maximum speed is $10\,\text{m s}^{-1}$ and its maximum acceleration is $10\,\text{m s}^{-2}$. Calculate the period and amplitude of the motion.

5 A particle is moving on a straight line with S.H.M. Its maximum speed is $12\,\text{m s}^{-1}$ and its maximum acceleration is $10\,\text{m s}^{-2}$. Calculate the speed of the particle when it is 2 m from the centre of oscillation.

6 A particle is moving in a straight line with S.H.M. of period 2π s about a centre O. When the particle is 0.5 m from O its speed is zero. Calculate the maximum speed of the particle and its speed when it is 0.25 m from O.

7 A particle P is moving in a straight line with S.H.M. of period
π s about a centre O. When $OP = 2$ m P has a speed of
$3 \, \text{m s}^{-1}$. Find the amplitude of the motion.
Given that when $t = 0$ P is passing through O, find an
expression for the displacement x m of P from O at any
subsequent time t s.

8 A particle P is moving in a straight line with S.H.M. centre O.
P passes through O with speed $6 \, \text{m s}^{-1}$ and performs 4
complete oscillations per second. Calculate the amplitude of
the motion and the distance OP when the magnitude of P's
acceleration is $30 \, \text{m s}^{-2}$.

9 A particle P is moving on a straight line with S.H.M. between
two points A and B. O is the mid-point of AB. When $OP = 1$ m,
P has speed $4 \, \text{m s}^{-1}$. When $OP = 0.5$ m P has speed $6 \, \text{m s}^{-1}$.
Determine the distance AB and the period of the motion.

10 A piston of mass 1.5 kg is moving with S.H.M. inside a
cylinder. It performs 20 complete oscillations per minute and
the distance between the extreme points of its motion is 2 m.
Calculate the maximum value of the kinetic energy possessed
by the piston during its motion, stating clearly any
assumptions you have made about the piston in order to
calculate this.

11 A marker buoy is oscillating in a vertical line with S.H.M. The
buoy takes 3 s for one oscillation and is seen to rise and fall
through a distance of 0.75 m. Choose a suitable model for the
marker buoy and hence calculate its maximum speed and the
time taken by the buoy to fall a distance of 0.5 m from its
highest point.

12 A particle is moving along the x-axis so that its displacement
x m from O at time t s is given by $x = 5 \sin 3t$. Prove that the
particle is moving with S.H.M. and state the amplitude and
period of the motion. Calculate (a) the maximum speed of the
particle (b) the value of t the first time its speed is $10 \, \text{m s}^{-1}$
(c) the value of t the first time $x = 2.5$.

13 A particle P is moving along the x-axis so that its
displacement x m from O at time t s is given by

$x = 2\sin(5t + \frac{1}{2})$. Prove that the particle is moving with S.H.M. and state the amplitude and period of the motion. Determine the value of x when $t = 0$ and the time when P passes through O for the first time.

14 Points A, O, B lie in that order on a straight line. A particle P is moving on the line with S.H.M. of period 3 s, amplitude 0.6 m and centre O. Given that OA is 0.3 m and OB is 0.4 m, calculate the time taken by P to move directly from A to B.

15 Points O, A, B lie in that order on a straight line. A particle P is moving on the line with S.H.M. of period 4 s, amplitude 0.5 m and centre O. OA is 0.1 m and OB is 0.3 m. When $t = 0$, P passes through B travelling in the direction OB. Calculate the time when P reaches A.

16 The water level in a harbour can be assumed to rise and fall with simple harmonic motion. On a certain day, low tide will occur at 12 noon and the depth of the water will be 3 m. The subsequent high tide will occur at 6.20 pm and the water will then be 15 m deep. A ship which needs 6 m depth of water wishes to use the harbour. Find, to the nearest minute, the earliest time it can enter the harbour on this day and the time by which it must leave.

3.4 Horizontal oscillations of a particle attached to the end of an elastic string or spring

Consider a particle P resting on a smooth horizontal surface and attached to one end of an elastic spring whose other end is fixed to a point O on the surface.

When there is no horizontal force applied to P the spring will be unstretched and so at its natural length l. However, if P is pulled aside in the direction OP and then released the spring will be stretched and so will exert a tension on P which will result in an acceleration.

Let the extension of the spring be x.

By Hooke's law the tension T in the spring is given by:

$$T = \frac{\lambda x}{l}$$

where l is the natural length and λ the modulus of the spring.

The acceleration $\ddot{x}$ of P will be directed *away* from the fixed point O of the spring as that is the direction of *increasing* x.

By Newton's second law, $F = ma$ and so:

$$-T = m\ddot{x}$$

where m is the mass of P.

Hence:
$$m\ddot{x} = -\frac{\lambda x}{l}$$

or:
$$\ddot{x} = -\frac{\lambda x}{ml} \qquad (1)$$

But λ, m and l all have positive, constant values. So equation (1) is of the form $\ddot{x} = -\omega^2 x$ and hence the particle is moving with S.H.M.

where $\omega^2 = \dfrac{\lambda}{ml}$

Since:
$$\text{period} = \frac{2\pi}{\omega}$$

the period of the particle's oscillations is $2\pi\sqrt{\left(\dfrac{ml}{\lambda}\right)}$.

The initial extension is also the maximum value of x. So the amplitude of the motion is the same as the initial extension.

The above work used Hooke's law to determine the tension in the spring. A *spring* will always have a tension or compression and so a particle attached to the end of a spring will perform complete oscillations whether it is pulled aside (spring initially stretched) or pushed aside (spring initially compressed) before release. However, for a particle attached to the end of a *string*, the motion will be oscillatory only while the string is taut. Once the string becomes slack there will be no tension and the particle will continue to move with constant speed until the string becomes taut again.

In the above diagram the string is stretched and P will accelerate.

Once the string returns to its natural length there is no tension and hence P has no acceleration but it has already acquired a velocity.

When P has moved a distance equal to twice the natural length of the string, the string will become taut again. The resulting tension will cause P to decelerate and oscillatory motion will once more take place.

Example 12

A small ball of mass $0.5\,$kg rests on a smooth horizontal table attached to one end of a piece of elastic of natural length $0.8\,$m and modulus $20\,$N. The other end of the elastic is fixed to a point on the table. The ball is pulled aside so that the elastic is extended by $0.2\,$m and then released.

(a) Describe the models you would use for the ball and elastic when investigating the resulting motion.
(b) Determine the speed of the ball when the elastic returns to its natural length.
(c) Calculate the time that elapses between the string becoming slack and becoming taut again. Hence calculate the time taken by the ball to return to its starting point for the first time.
(d) Describe briefly how your model could be refined.

(a) The ball is small compared with the length of the elastic and so can be modelled as a particle. The elastic can be modelled as a light elastic string.

(b) Consider the ball when the elastic is stretched $x\,$m. The total length of the elastic is then $(0.8 + x)\,$m. Let the tension in the elastic be $T\,$N and the acceleration of the ball be $\ddot{x}\,$m s^{-2} as shown.

Using Hooke's law. $T = \dfrac{\lambda x}{l}$ with $\lambda = 20\,$N and $l = 0.8\,$m gives:

$$T = \frac{20x}{0.8}$$

$$T = 25x \qquad\qquad (1)$$

Using Newton's second law $F = ma$ with $m = 0.5\,$kg gives:

$$T = -0.5\ddot{x} \qquad\qquad (2)$$

Combining equations (1) and (2) gives:

$$25x = -0.5\ddot{x}$$

So:

$$\ddot{x} = -50x$$

Comparing this with the standard S.H.M. equation $\ddot{x} = -\omega^2 x$ shows $\omega = \sqrt{50}$.

The amplitude of the motion is the same as the initial extension of the string. So the amplitude is 0.2 m.

To determine the speed of the ball when the string becomes slack you must use the standard S.H.M. equation:

$$v^2 = \omega^2(a^2 - x^2)$$

with $\omega = \sqrt{50}$, $a = 0.2$ and $x = 0$.

So:

$$v^2 = 50(0.2^2 - 0)$$

$$v = 1.414$$

The speed of the ball when the elastic returns to its natural length is $1.41\,\mathrm{m\,s^{-1}}$.

(c)

The ball will move from P to Q while the string is slack. As the natural length of the string is 0.8 m, distance $PQ = 2 \times 0.8\,\mathrm{m} = 1.6\,\mathrm{m}$

The speed of the ball is $1.414\,\mathrm{m\,s^{-1}}$.

As:

$$\text{time} = \frac{\text{distance}}{\text{speed}}$$

the time is $\left(\dfrac{1.6}{1.414}\right)\mathrm{s} = 1.131\,\mathrm{s}$

The string is slack for 1.13 s.

To calculate the time taken by the ball to return to its starting point you must include the time for which the ball is oscillating. From (b) above $\omega = \sqrt{50}$.

So the period of the oscillation $= \dfrac{2\pi}{\omega} = \dfrac{2\pi}{\sqrt{50}}$

The whole motion consists of half an oscillation at each end of the ball's path and the constant speed section in the middle. The middle section is covered twice.

So: total time $=$ period of oscillation $+ 2 \times$ time from P to Q

$$= \frac{2\pi}{\sqrt{50}} + 2 \times 1.131$$

$$= 3.150$$

The particle will return to its starting point after $3.15\,\text{s}$.

(d) The table is unlikely to be perfectly smooth. The model could therefore be refined by including the frictional force between the table and the ball.

Example 13

A particle P of mass $0.5\,\text{kg}$ is attached to the ends of two identical springs of natural length $0.5\,\text{m}$ and modulus of elasticity $10\,\text{N}$. The free ends of the springs are attached to two points A and B which are $2\,\text{m}$ apart on a smooth horizontal surface. P is held at point C where $AC = 1.2\,\text{m}$ and A, B and C are collinear. P is released from rest. Show that the subsequent motion is simple harmonic. Calculate the maximum speed of P and hence its maximum kinetic energy.

The springs are identical, so P is in equilibrium at O, the mid-point of AB. This will be the centre of the oscillation.

When P is at a distance $x\,\text{m}$ from its equilibrium position as in the diagram, its acceleration will be $\ddot{x}$ in the direction away from O.

The extension in AP is $(1 + x - 0.5) = (0.5 + x)\,\text{m}$ and the extension in PB is $(1 - x - 0.5) = (0.5 - x)\,\text{m}$.

Let the tension in AP be T_A and the tension in PB be T_B.

Using Hooke's law: $\qquad T = \dfrac{\lambda x}{l}$

gives: $\qquad T_A = \dfrac{10(0.5 + x)}{0.5} \qquad T_B = \dfrac{10(0.5 - x)}{0.5}$

Using the equation of motion:

$$F = ma$$

gives: $\qquad T_B - T_A = 0.5\,\ddot{x}$

$$\frac{10(0.5 - x)}{0.5} - \frac{10(0.5 + x)}{0.5} = 0.5\,\ddot{x}$$

$$0.5\,\ddot{x} = -2 \times \frac{10x}{0.5}$$

$$\ddot{x} = -\frac{20x}{0.5^2} = -80x$$

As this equation is of the form $\ddot{x} = -\omega^2 x$, the particle is moving with simple harmonic motion.

From $\ddot{x} = -80x$ it follows that $\omega^2 = 80$.

The particle is at rest when $x = (1.2 - 1) = 0.2$ m. So the amplitude, a, of the motion is 0.2 m.

Using: $$v^2 = \omega^2(a^2 - x^2)$$

gives: $$v^2 = 80(0.2^2 - x^2)$$

The maximum speed will occur when $x = 0$.

Hence: $$v_{max}^2 = 80 \times 0.2^2$$

$$v_{max} = 1.79 \, \text{m s}^{-1}$$

The maximum kinetic energy is obtained by using $\frac{1}{2}mv^2$ with the maximum speed.

Maximum K.E. $= \frac{1}{2} \times 0.5 v_{max}^2 = \frac{1}{2} \times 0.5 \times 80 \times 0.2^2 = 0.8 \, \text{J}$.

Exercise 3D

1 A particle P of mass 0.2 kg is attached by an elastic string of modulus 15 N and natural length 1 m to a point A of the smooth horizontal surface on which P rests. P receives an impulse of magnitude 0.5 N s in the direction AP. Show that while the string is taut the motion of P is simple harmonic and calculate the period and amplitude of the motion.

2 A particle P of mass 1.2 kg is attached by an elastic spring of modulus 100 N and natural length 0.5 m to a point A on the smooth horizontal table on which P rests. P receives an impulse of magnitude 2 N s in the direction AP. Given that t seconds after the blow the displacement of P from its initial position is x m, find an equation for x in terms of t and calculate the maximum acceleration of P.

3 A small ball of mass 0.7 kg is attached to one end of a light elastic spring of natural length 0.6 m and modulus 84 N. The other end of the spring is fixed to a point on the smooth horizontal surface on which the ball rests. The ball is pushed aside so that the spring has length 0.4 m and then released. Calculate the period of the resulting oscillations and the maximum speed of the ball.

4   A small ball of mass 0.3 kg rests on a smooth horizontal table
    attached to one end of a piece of elastic of natural length 1 m
    and modulus 48 N. The other end of the elastic is fixed to a
    point on the table. The ball is pulled aside so that the elastic
    measures 1.3 m and then released.
    (a) Choose suitable models for the ball, elastic and table so
    that you can investigate the motion of the ball.
    (b) Determine the speed of the ball when the elastic is slack.
    (c) Calculate the time taken by the ball to return to its
    starting point for the first time.
    (d) Your models allow a force (other than gravity) to be
    neglected. Explain briefly why this is possible. If this force
    were included, how would your answer to part (b) be
    affected?

5   A particle of mass 0.75 kg rests on a smooth horizontal
    surface attached to one end of a light elastic string of natural
    length 1.5 m and modulus 20 N. The other end of the string
    is attached to a point on the surface. The particle is pulled
    aside so that the string measures 2 m and released. State the
    amplitude of the resulting oscillation and calculate the speed
    of the particle when the string becomes slack.
    Before the string becomes taut again the particle hits a
    vertical surface which is at right angles to the particle's
    direction of motion. The particle rebounds with speed half
    that with which it hits the surface. Calculate the amplitude
    of the oscillation which takes place when the string is once
    more taut.

6   A piston of mass 2 kg is moving in a smooth cylinder which is
    fixed with its axis horizontal. The piston is attached to the end
    of the cylinder by a spring of natural length 40 cm and
    modulus 500 N. The piston is pushed aside until the spring is
    compressed to a length of 30 cm and then released.
    (a) Choose suitable models for the spring and piston stating
    clearly any assumptions you make.
    (b) Determine the period of the resulting oscillations and the
    maximum speed of the piston.
    (c) Describe briefly how your model could be refined.

7 A small ball of mass 1.2 kg is attached to an elastic string of unstretched length 0.75 m and modulus 60 N. The other end of the string is fixed to a point of the smooth horizontal floor on which the ball rests. The ball is pulled aside until the string measures 0.9 m and released. Calculate the speed of the ball when the string becomes slack.

When the ball has travelled a distance 0.6 m from its point of release, the floor becomes rough. Given that the coefficient of friction between the ball and the floor is 0.3, determine the time for which the ball is moving.

8 Two identical elastic strings of natural length 0.8 m and modulus 4 N are attached to a particle P of mass 0.75 kg. Their free ends are attached to two points A and B on a smooth horizontal surface where $AB = 3$ m. P is held at a point C, where $AC = 1$ m and A, C and B are collinear, and released. Show that the subsequent motion is simple harmonic and give the amplitude and period of the motion. Calculate the maximum speed of P and hence its maximum kinetic energy.

9 A particle P of mass 1 kg is attached to two identical light springs of natural length 0.75 m and modulus of elasticity 8 N. The free ends of the springs are attached to two points A and B which are 2.5 m apart on a smooth horizontal surface. P is held at point C where $AC = 0.8$ m and A, B and C are collinear. P is released from rest. Show that the subsequent motion is simple harmonic. Calculate the maximum speed of P and hence its maximum kinetic energy.

10 A particle P of mass m is attached to two identical light strings of natural length l and modulus of elasticity $5mg$. The free ends of the strings are attached to two fixed points A and B which are $4l$ apart on a smooth horizontal surface. P is released from rest from the point C where $AC = l$ and A, B and C are collinear. Show that the subsequent motion is simple harmonic and find the amplitude and period of the motion. Find the speed of P when $AP = \dfrac{3l}{2}$.

3.5 Vertical oscillations of a particle attached to the end of an elastic string or spring

Consider a particle which is hanging in equilibrium attached to the end of an elastic string or spring, the other end of which is fixed. If the particle is pulled downwards and released it will oscillate in a vertical line about its equilibrium position.

Let the particle be a distance x below its equilibrium position. The acceleration of P will be $\ddot{x}$ *away* from the equilibrium position as that is the direction of *increasing* x. By Hooke's law the tension in the string or spring is given by $T = \dfrac{\lambda}{l} \times$ extension where l is the natural length of the string or spring and λ is the modulus. Let the extension in the string or spring when the particle is in equilibrium be e. Then the total extension in the string or spring is $x + e$.

Let the tension when the particle is in equilibrium be T_0. In the equilibrium position resolving vertically gives:

$$T_0 = mg$$

Hooke's law gives:

$$T_0 = \frac{\lambda e}{l}$$

And so:

$$\frac{\lambda e}{l} = mg$$

$$e = \frac{mgl}{\lambda}$$

Hence the total extension when the particle is displaced a distance x from the equilibrium position is $e + x = \dfrac{mgl}{\lambda} + x$.

The tension, T, at this point is given by:

$$T = \frac{\lambda}{l}\left(\frac{mgl}{\lambda} + x\right) = mg + \frac{\lambda x}{l}$$

Using Newton's second law, $F = ma$, gives:

$$mg - T = m\ddot{x}$$

and so:

$$mg - \left(mg + \frac{\lambda x}{l}\right) = m\ddot{x}$$

$$m\ddot{x} = -\frac{\lambda x}{l}$$

$$\ddot{x} = -\frac{\lambda x}{ml}$$

As in the case with horizontal oscillations, the particle is moving with S.H.M. of period $2\pi\sqrt{\left(\dfrac{ml}{\lambda}\right)}$.

As in the horizontal case, a particle attached to the end of a spring will perform complete oscillations. In the case of a string however, the S.H.M. equation will only hold while the string is taut. That is, the particle must not rise above its equilibrium position a distance greater than its equilibrium extension e. The greatest distance that can be reached above the equilibrium position while moving with S.H.M. is the amplitude. So the amplitude must be less than or equal to the equilibrium extension e. If this is not the case the string will become slack when $x = -e$ and the particle will then move freely under gravity until it falls back to the position where $x = -e$ once more.

To summarise:

A particle which is hanging in equilibrium attached to the end of an elastic *spring* and is displaced vertically from its equilibrium position and then released will move with S.H.M. performing complete oscillations.

A particle attached to the end of an elastic *string* and similarly displaced and released will move with S.H.M. performing complete oscillations if the initial displacement is less than the equilibrium extension e. When the string becomes slack the particle will move freely under gravity.

Example 14

A piston of mass 1.5 kg moves in a smooth cylinder which is fixed
with its axis vertical. It is attached to the base of the cylinder by a
spring of natural length 20 cm and modulus 200 N as shown in the
diagram.

(a) Choose suitable models for the spring and piston, stating clearly
any assumptions you make.
(b) Determine the compression of the spring when the piston rests
in equilibrium.

The piston is now pushed downwards a further 0.5 cm and
released.

(c) Calculate the period of the resulting oscillations and the
maximum speed of the piston.

(a) The mass of the spring can be assumed to be small compared
with that of the piston. The spring can therefore be modelled as a
light elastic spring.

The piston is attached to the end of the spring. Assuming the piston
is a uniform solid and the spring is attached to the centre of the base
of the piston, the weight of the piston will act along the line of the
spring. Thus the piston will have the same effect on the spring as
would a particle of mass equal to that of the piston attached to the
end of the spring. In other words, the piston can be modelled as a
particle.

(b)

Consider the piston when it is in equilibrium. A diagram of the
model is shown above.

Let the thrust in the spring be T_0 N and the equilibrium compression
be e m.

Resolving vertically gives:

$$T_0 = 1.5g$$

Using Hooke's law:

$$T = \frac{\lambda \times \text{extension}}{\text{natural length}}$$

with $T = 1.5g$ N, $\lambda = 200$ N, extension e m and natural length 20 cm $= 0.2$ m gives:

$$1.5g = \frac{200e}{0.2}$$

$$e = \frac{1.5 \times 9.8 \times 0.2}{200}$$

$$e = 0.0147$$

The compression when the piston is in equilibrium is 0.0147 m or 1.47 cm.

(c) Consider the piston when it is a distance x below its equilibrium position. The acceleration will be $\ddot{x}$ m s^{-2} downwards and the compression of the spring will be $(0.0147 + x)$ m.

Using Newton's second law $F = ma$ with $m = 1.5$ kg gives:

$$1.5g - T = 1.5\ddot{x} \qquad (1)$$

Using Hooke's law:

$$T = \frac{\lambda \times \text{extension}}{\text{natural length}}$$

with $\lambda = 200$ N, extension $= (0.0147 + x)$ m and natural length 0.2 m gives:

$$T = \frac{200(0.0147 + x)}{0.2}$$

Substituting this expression for T in equation (1) gives:

$$1.5 \times 9.8 - \frac{200(0.0147 + x)}{0.2} = 1.5\ddot{x}$$

$$14.7 - 14.7 - 1000x = 1.5\ddot{x}$$

$$\ddot{x} = -\frac{1000}{1.5} x$$

or:

$$\ddot{x} = -\frac{2000}{3} x \qquad (2)$$

Comparing equation (2) with the standard S.H.M equation $\ddot{x} = -\omega^2 x$ gives:

$$\omega^2 = \frac{2000}{3}$$

So:
$$\omega = \sqrt{\left(\frac{2000}{3}\right)}$$

Since:
$$\text{period} = \frac{2\pi}{\omega}$$

this gives:
$$\text{period} = 2\pi\sqrt{\left(\frac{3}{2000}\right)} = 0.2433$$

The period of the resulting oscillations is $0.243\,\text{s}$.

The speed of the piston is found by using the standard formula:
$$v^2 = \omega^2(a^2 - x^2)$$

The maximum speed will occur when x^2 is minimum; that is when $x = 0$.

In the equation a is the maximum distance of the piston from its equilibrium position. As the piston was pushed downwards until it was $0.5\,\text{cm}$ below its equilibrium position, $a = 0.5\,\text{cm} = 0.005\,\text{m}$.

Using:
$$v^2 = \omega^2(a^2 - x^2)$$

with $a = 0.005$, $\omega^2 = \dfrac{2000}{3}$ and $x = 0$ gives:

$$v^2 = \frac{2000}{3}(0.005^2 - 0)$$
$$v = 0.1290$$

The maximum speed is $0.129\,\text{m s}^{-1}$.

Example 15

A particle P of mass $0.1\,\text{kg}$ is hanging in equilibrium attached to one end of a light elastic string of natural length $0.5\,\text{m}$ and modulus $5\,\text{N}$. The other end of the string is fixed. P is pulled downwards a further $0.2\,\text{m}$ and released. Calculate the time that elapses before the string becomes slack and the further time before the string is once more taut.

First the equilibrium level must be determined. Let the extension when the particle is in equilibrium be $e\,\text{m}$ and the tension in the string be $T_0\,\text{N}$.

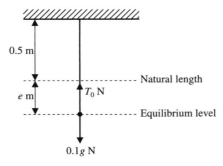

For the particle in equilibrium, resolving vertically gives:

$$T_0 = 0.1g$$

Using Hooke's law:

$$T = \frac{\lambda \times \text{extension}}{\text{natural length}}$$

with $\lambda = 5\,\text{N}$ and natural length $0.5\,\text{m}$ gives:

$$T_0 = \frac{5e}{0.5} = 10e$$

So:

$$10e = 0.1g$$

$$e = 0.01 \times 9.8 = 0.098$$

The particle is pulled downwards a further $0.2\,\text{m}$ and released. The motion, while the string is taut, will be simple harmonic with amplitude $0.2\,\text{m}$.

To use the standard S.H.M. formulae, ω must be known. Consider P when it is a distance x below its equilibrium position. The acceleration will be $\ddot{x}$ downwards as shown.

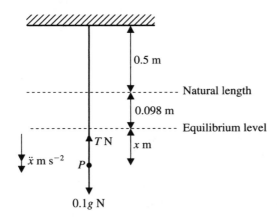

Using Newton's second law, $F = ma$, gives:

$$0.1g - T = 0.1\ddot{x} \qquad (1)$$

Using Hooke's law:

$$T = \frac{\lambda \times \text{extension}}{\text{natural length}}$$

with $\lambda = 5\,\text{N}$, extension $= (0.098 + x)\,\text{m}$, natural length $0.5\,\text{m}$ gives:

$$T = \frac{5(0.098 + x)}{0.5}$$

$$T = 10(0.098 + x)$$

And substituting this expression for T in equation (1) gives:

$$0.1 \times 9.8 - 10(0.098 + x) = 0.1\ddot{x}$$

$$0.98 - 0.98 - 10x = 0.1\ddot{x}$$

$$\ddot{x} = -\frac{10x}{0.1}$$

$$\ddot{x} = -100x$$

Comparing this with the standard S.H.M. equation $\ddot{x} = -\omega^2 x$ shows that $\omega = 10$.

To find the time that elapses before the string becomes slack use the equation

$$x = a\sin(\omega t + \alpha)$$

Substituting $x = a$ and $t = 0$ gives:

$$a = a\sin\alpha$$

So: $$\sin\alpha = 1$$

and: $$\alpha = \frac{\pi}{2}$$

Therefore: $$x = a\sin\left(\omega t + \frac{\pi}{2}\right)$$

or: $$x = a\cos\omega t$$

From above, $a = 0.2\,\mathrm{m}$ and $\omega = 10$, so:

$$x = 0.2\cos 10t$$

The string will become slack when there is no extension, that is when $x = -0.098\,\mathrm{m}$

So: $$-0.098 = 0.2\cos 10t$$

$$-0.49 = \cos 10t \qquad (2)$$

When the string *first* becomes slack $\frac{\pi}{2} < 10t < \pi$

So the required solution for equation (2) is

$$10t = 2.082$$

$$t = 0.2082$$

The string will become slack after 0.208 s.

While the string is slack the particle moves vertically under gravity alone. To calculate the time for which this happens you must first calculate the speed of the particle at the moment the string becomes slack.

Using $v^2 = \omega^2(a^2 - x^2)$ with $x = -0.098\,\mathrm{m}$, $\omega = 10$ and $a = 0.2\,\mathrm{m}$ gives:

$$v^2 = 10^2(0.2^2 - (-0.098)^2)$$

$$v = 1.743$$

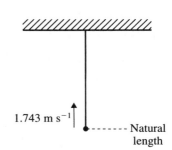

1.743 m s^{-1} ⌐ ----- Natural
length

For the motion under gravity the known quantities are $u = 1.743\,\mathrm{m\,s^{-1}}$ and $a = -9.8\,\mathrm{m\,s^{-2}}$. The string will become taut once more when $s = 0\,\mathrm{m}$.

Using:
$$s = ut + \tfrac{1}{2}at^2$$

gives:
$$0 = 1.743t - \tfrac{1}{2} \times 9.8t^2$$
$$0 = (1.743 - 4.9t)t$$

The solutions are: $t = 0$ (the moment the string becomes slack)

or:
$$t = \frac{1.743}{4.9}$$

$t = 0.3557$ (the moment the string becomes taut once more)

The further time that elapses before the string becomes taut once more is $0.356\,\mathrm{s}$.

Exercise 3E

Whenever a numerical value of g is required, take $g = 9.8\,\mathrm{m\,s^{-2}}$.

1 A particle of mass 2 kg is attached to the free end of an elastic spring of natural length 2 m and modulus 80 N which is hanging vertically. Determine the extension of the spring when the particle is hanging in equilibrium. The particle is pulled downwards a further 0.5 m and released. Calculate the period of the resulting oscillations and the maximum speed of the particle.

2 A particle of mass 0.5 kg is hanging in equilibrium attached to the free end of an elastic spring of natural length 0.8 m and modulus 60 N which is hanging vertically. The particle is pulled downwards and released. Calculate the period of the resulting oscillations. Given that the particle passes through its equilibrium position with speed $2\,\mathrm{m\,s^{-1}}$, calculate the distance below the equilibrium position of the point from which the particle was released.

3 A particle of mass 1.5 kg is hanging in equilibrium attached to the free end of an elastic spring of natural length 1.2 m which is hanging vertically. The particle receives an impulse of 6 N s parallel to the spring. Show that the subsequent motion is simple harmonic.

Given that the period of the motion is $\frac{2\pi}{5}$ calculate the modulus of the spring and the amplitude of the motion.

4 A light elastic spring has one end A fixed and hangs vertically with a particle P of mass 0.2 kg attached to its free end. The particle is pulled downwards from its equilibrium position and released. P subsequently moves with simple harmonic motion of period $\frac{\pi}{3}$ s and the maximum and minimum distances of P below A are 1.5 m and 1 m. Calculate (a) the distance below A of the centre of the oscillation (b) the amplitude of the oscillation (c) the maximum speed of P.

5 A child's toy consists of a 'man' attached to a spring which is fixed to a board. The mass of the man is 0.25 kg. When in equilibrium with the spring vertical as shown in the diagram, the spring is compressed by 0.02 m. The man is pushed downwards a further 0.02 m and then released from rest. Choose suitable models for the man and spring, stating clearly the assumptions you make, and hence calculate the amplitude and period of the resulting oscillations.

6 A piston of mass 2.6 kg moves in a smooth cylinder which is fixed with its axis vertical. It is attached to the base of the cylinder by a spring of natural length 15 cm and modulus 400 N. Assuming that the spring is attached to the centres of the bases of the piston and cylinder, choose suitable models for the piston and spring stating clearly any further assumptions you make.
The piston is released from rest at a point where the spring has length 12 cm. Calculate the period of the resulting oscillations and the maximum speed of the piston.
State briefly how the model could be refined to enable calculation of more accurate solutions (do not perform any calculations).

7 A particle P of mass 0.3 kg is hanging in equilibrium attached to one end of a light elastic string of natural length 0.6 m and modulus 9 N. P is pulled downwards a further 0.4 m and released. Calculate the time that elapses before the string

becomes slack and the further time before the string is once more taut.

8 A particle P of mass 2 kg is hanging in equilibrium attached to one end of a light elastic string of natural length 1.2 m and modulus 48 N. P is pulled downwards a distance 0.5 m and released. When P has risen a distance 0.3 m the string is cut. Calculate the greatest height P reaches above its equilibrium position and the time taken to reach that greatest height.

9 A rock climber of mass 80 kg is attached to one end of a flexible rope of natural length 20 m and modulus 45 000 N. She slips from an overhang which is 10 m vertically below the point to which the other end of the rope is fixed.
(a) How could you model the climber and the rope to investigate her subsequent motion? State clearly any assumptions you would make about the rope and climber. Assuming that the climber falls vertically without hitting the rock face, calculate (b) the speed of the climber when the rope becomes taut (c) the maximum distance of the climber below the point where she slipped (d) the time from slipping to reaching her lowest point.

10 A particle P of mass 2 kg is attached to the mid-point of a light elastic string of natural length 1.5 m and modulus of elasticity 20 N. One end of the string is attached to point A of a ceiling and the other end is attached to point B on the floor 3 m vertically below A. Given that P is in equilibrium,
(a) calculate the length of AP.
The particle is pulled vertically down a distance 0.1 m from its equilibrium position and released from rest.
(b) Show that the subsequent motion is simple harmonic.
(c) Find the amplitude and period of the motion.

11 A light string has its two ends fastened together to form a loop of natural length $2l$ and modulus of elasticity mg. The string is placed over two smooth fixed pegs A and B where AB is horizontal and of length $2l$. A particle P of mass $2m$ is attached to the mid-point of the lower part of the loop and hangs in equilibrium vertically below C, the mid-point of AB.

(a) Explain why the extension of the string is equal to twice the length of AP.

(b) Show that the distance CP is l.

P is projected vertically downwards from its equilibrium position.

(c) Show that, providing the string does not break and P does not reach the level of AB, the subsequent motion is simple harmonic.

(d) Find the period of the motion.

Exercise 3F   Mixed questions

Whenever a numerical value of g is required take $g = 9.8\,\mathrm{m\,s^{-2}}$.

1 A particle P is moving with simple harmonic motion in a straight line between two fixed points which are $5\,\mathrm{m}$ apart. The particle completes two oscillations every second. Find the maximum speed of the particle.

2 A particle P of mass $0.25\,\mathrm{kg}$ moves along the positive x-axis under the action of a force directed towards the origin of magnitude $\dfrac{4}{x^3}\,\mathrm{N}$ where x metres is the distance of P from the origin. P passes through the point where $x = 1$ with speed $8\,\mathrm{m\,s^{-1}}$. Find:

(a) the distance of P from the origin when its speed is $7\,\mathrm{m\,s^{-1}}$

(b) the work done by the force while reducing P's speed from $8\,\mathrm{m\,s^{-1}}$ to $7\,\mathrm{m\,s^{-1}}$.

3 A particle P of mass m is attached to one end of a light elastic string of natural length $2l$ and modulus of elasticity $3mg$. The other end of the string is attached to a fixed point A on a ceiling. The particle hangs in equilibrium.

(a) Find the length of the string.

The particle is raised so that the string is still vertical and $AP = \dfrac{7l}{3}$. The particle is released from rest.

(b) Show that the subsequent motion is simple harmonic.

(c) Find the period of the motion.

(d) Find the speed of P as it passes through its equilibrium position.

4 A particle P of mass $2\,\text{kg}$ moves along the positive x-axis.
   When P's displacement from the origin is $x\,\text{m}$, the force acting
   on P is $10x\,\text{N}$ and its velocity is $v\,\text{m}\,\text{s}^{-1}$. When $x = 5$, $v = 8$.
   (a) Find an expression for v^2 in terms of x^2.
   (b) Find the distance of P from O when $v = 17$.

5 A light elastic string of natural length $75\,\text{cm}$ is hanging vertically
   with one end fixed to a point of a ceiling. A particle P of mass
   $0.5\,\text{kg}$ is attached to the other end of the string. When P is
   hanging in equilibrium the length of the string is $90\,\text{cm}$.
   (a) Find the modulus of elasticity of the string.
   The string is now placed on a smooth horizontal surface and a
   second, identical string is attached to P. The two free ends are
   attached to fixed points A and B of the surface where
   $AB = 2.5\,\text{m}$. P is held at rest at point C of the table, between
   A and B where $AC = 1\,\text{m}$. P is released from rest.
   (b) Show that P moves with simple harmonic motion.
   (c) Find the period of the motion.
   (d) Find the maximum value of the kinetic energy of P.

6 A particle P moves along the x-axis in such a way that at time
   t seconds its distance x metres from the origin O is given by

   $$x = 6\cos\left(\frac{\pi t}{4}\right)$$

   (a) Show that P is moving with simple harmonic motion.
   (b) Write down the period and the amplitude of the motion.
   (c) Find the maximum speed of P.
   The points A and B are such that $OA = OB = 3\,\text{m}$.
   (d) Find the time taken by P to travel directly from A to B.

7 A particle P of mass $0.5\,\text{kg}$ moves in a straight line. At time t
   seconds the acceleration of P is $e^{3t}\,\text{m}\,\text{s}^{-2}$, its speed is $v\,\text{m}\,\text{s}^{-1}$
   and its displacement from the origin is $x\,\text{m}$. When $t = 0$, P is
   at rest at the origin.
   (a) Find the magnitude of the force acting on P when $t = 2$.
   (b) Show that $v = \frac{1}{3}\left(e^{3t} - 1\right)$.
   (c) Find, in terms of t, an expression for x.

8 The amplitude of oscillation of a particle describing simple
   harmonic motion is $3\,\text{m}$. When the particle is $1\,\text{m}$ from the
   centre of the oscillation its speed is $4\,\text{m}\,\text{s}^{-1}$. Find the periodic
   time of the motion.

SUMMARY OF KEY POINTS

1 For a particle of mass m moving in a straight line under the influence of a force $F = F(t)$

$$m\frac{\mathrm{d}v}{\mathrm{d}t} = F(t)$$

2 For a particle of mass m moving in a straight line under the influence of a force $F = G(x)$

$$m\frac{\mathrm{d}}{\mathrm{d}x}\left(\frac{1}{2}v^2\right) = G(x)$$

3 The impulse of a variable force $F(t)$ acting over the time interval t_1 to t_2 is:

$$\int_{t_1}^{t_2} F(t)\,\mathrm{d}t$$

4 The work done by a variable force $G(x)$ which moves its point of application from x_1 to x_2 is:

$$\int_{x_1}^{x_2} G(x)\mathrm{d}x$$

5 A particle which moves on a straight line so that its acceleration is always towards a fixed point O in the line and is proportional to its displacement from O is said to move with simple harmonic motion or S.H.M.

This is written as: $\ddot{x} = -\omega^2 x$

The maximum displacement of the particle from O is the amplitude, a, of the motion.

The period of the motion is $\dfrac{2\pi}{\omega}$

The speed, v, at any point is given by

$$v^2 = \omega^2(a^2 - x^2)$$

and $x = a\sin\omega t$ if $x = 0$ when $t = 0$
 $x = a\cos\omega t$ if $x = a$ when $t = 0$
 $x = a\sin(\omega t + \alpha)$ if x is at some other point when $t = 0$

6 A particle attached to the end of an elastic spring which is displaced from its equilibrium position and then allowed to move under the influence of the tension or thrust only will move with S.H.M.

If the particle is attached to a string, the motion will be simple harmonic only while the string is taut.

Review exercise 1

Whenever a numerical value of g or G is required take
$g = 9.8\,\mathrm{m\,s^{-2}}$ and $G = 6.67 \times 10^{-11}\,\mathrm{kg^{-1}\,m^3\,s^{-2}}$.

1 A particle moves along Ox in the positive direction. At time
   $t = 0$, its speed is $8\,\mathrm{m\,s^{-1}}$. At time t s, its acceleration is
   $3e^{2t}\,\mathrm{m\,s^{-2}}$ in the positive direction. Calculate, in metres to
   3 significant figures, the distance the particle moves in the first
   2 seconds.   [E]

2 A particle travels along a straight line which passes through A
   and B. During the motion the velocity of the particle is
   $6t + t^2$ metres per second, where t seconds is the time
   measured from a certain instant. The particle passes through A
   when $t = 2$ and through B when $t = 5$. Find, in terms of t, the
   acceleration of the particle and its distance from A at any
   instant during the motion. Calculate, also, the distance of B
   from A.   [E]

3 A particle starts with speed $20\,\mathrm{m\,s^{-1}}$ and moves in a straight
   line. The particle is subjected to a resistance which produces a
   retardation which is initially $5\,\mathrm{m\,s^{-2}}$ and which increases
   uniformly with the distance moved, having a value of $11\,\mathrm{m\,s^{-2}}$
   when the particle has moved a distance of $12\,\mathrm{m}$. Given that the
   particle has speed $v\,\mathrm{m\,s^{-1}}$ when it has moved a distance of $x\,\mathrm{m}$,
   show that, while the particle is in motion,

   $$v\frac{\mathrm{d}v}{\mathrm{d}x} = -(5 + \tfrac{1}{2}x)$$

   Hence, or otherwise, calculate the distance moved by the
   particle in coming to rest.   [E]

4 A particle moving along a straight line starts at time $t = 0$ seconds with a velocity $4\,\mathrm{m\,s^{-1}}$. At any subsequent time t seconds the acceleration of the particle is $(6t - 8)\,\mathrm{m\,s^{-2}}$. Find

(a) the distance the particle moves before first coming to instantaneous rest,

(b) the total time T seconds taken by the particle to return to the starting point,

(c) the greatest speed of the particle for $0 \leqslant t \leqslant T$.   [E]

5 At time t seconds a particle moving on a straight line with velocity $v\,\mathrm{m\,s^{-1}}$ experiences an acceleration $kt\,\mathrm{m\,s^{-2}}$, where k is a positive constant. When $t = 0$, $v = 4$ and when $t = 10$, $v = 12$.

(a) Show that $k = \frac{4}{25}$.

(b) Find the distance covered by the particle in the interval $5 \leqslant t \leqslant 10$.   [E]

6 A particle moves on the positive x-axis. The particle is moving towards the origin O when it passes through the point A, where $x = 2a$, with speed $\sqrt{(k/a)}$, where k is constant. Given that the particle experiences an acceleration $k/(2x^2) + k/(4a^2)$ in a direction away from O, show that it comes instantaneously to rest at a point B, where $x = a$. Immediately the particle reaches B the acceleration changes to $k/(2x^2) - k/(4a^2)$ in a direction away from O. Show that the particle next comes instantaneously to rest at A.   [E]

7 A particle moves in a straight line so that its distance, s metres, from the origin after time t seconds is given by $s = t + \sin 2t$. Find expressions for

(a) the velocity $v\,\mathrm{m\,s^{-1}}$

(b) the acceleration, $f\,\mathrm{m\,s^{-2}}$

Hence show that

$$f^2 = 16 - 4(v - 1)^2$$ [E]

8 A particle P starts from rest at a fixed point O and moves on a horizontal plane in a straight line Ox. At time t s after leaving O the acceleration of P is $(15 + 4t - 3t^2)\,\mathrm{m\,s^{-2}}$. The particle comes to instantaneous rest at the point A.

(a) Show that P takes 5 s to reach A from O.

(b) Calculate the distance between O and A, giving your answer to the nearest m.

The mass of P is 0.2 kg. As P moves from O to A during the interval $0 \leqslant t \leqslant 5$, calculate

(c) the greatest kinetic energy of P, giving your answer in J,

(d) the greatest value of the magnitude of the force exerted on P, giving your answer in newtons.   [E]

9   A ball of mass 0.1 kg falls vertically from rest in a tank of liquid. The ball has speed v m s^{-1} at time t seconds and the resistance to motion is $0.98(1 - e^{t})$ N.

(a) Show that the acceleration of B at time t seconds is:
$$9.8e^{t} \, \text{m s}^{-2}$$

(b) Find v in terms of t.

(c) State any assumptions you made about the ball when modelling its motion.

10   A particle P of mass m is dropped from a height h above the surface of the liquid in a container and falls freely under gravity until it enters the liquid. As P enters the liquid, $\frac{1}{4}$ of its kinetic energy is lost. At time t after entering the liquid P has fallen a vertical distance x through the liquid and the resistance to motion is $3mx^2$. Calculate the speed of P when it has just entered the liquid and hence show that its speed at time t later is given by:
$$v^2 = 2gx - 2x^3 + \tfrac{3}{2}gh$$

11   A packet of mass 2 kg is dropped from the top of a tower which is 60 m high.

(a) Calculate the time taken by the packet to reach the ground, stating clearly any assumptions you make about the packet and the forces acting on it.

(b) The model can be refined by including the resistive force acting on the packet. Given that this resistive force has magnitude $19.6(1 - e^{\frac{-kt}{2}})$ N at time t seconds after the packet was dropped, find an expression for the speed v m s^{-1} of the packet at time t seconds.

12 A particle P of mass $0.5\,\text{kg}$ moves away from the origin O
along the positive x-axis under the action of a force directed
away from O. At time t seconds after P leaves O the force has
magnitude $(10 - 2t)\,\text{N}$. When $t = 0$ P has speed $5\,\text{m}\,\text{s}^{-1}$ and
when $t = 3$ the force is removed. Calculate:
(a) the magnitude of the impulse given to the particle by the
force between $t = 0$ and $t = 3$,
(b) the final speed of the particle.

13 A particle of mass $0.4\,\text{kg}$ is initially at rest on a smooth
horizontal surface. A horizontal force of magnitude
$(2x + 1)\,\text{N}$, where $x\,\text{m}$ is the distance of the particle from its
initial position, acts on the particle causing it to move in a
straight line. Calculate:
(a) the work done by the force in moving the particle $3\,\text{m}$
from its starting point,
(b) the speed of the particle when $x = 3$.

14 A particle P of mass $0.4\,\text{kg}$ moves along the positive x-axis
under the action of a single force of magnitude $\dfrac{12}{x^3}$ newtons,
where x metres is the distance of P from the origin O. The
force acts along the x-axis in the direction of x increasing. The
particle is released from rest at the point where $x = 2$. In the
subsequent motion, when P is x metres from O, its speed is v
metres per second. Show that

$$v^2 = \frac{15}{2}\left(1 - \frac{4}{x^2}\right)$$ [E]

15 A particle P, of mass $0.2\,\text{kg}$, moves in a straight line through a
fixed point O. At time t seconds after passing through O, the
distance of P from O is x metres, the velocity of P is $v\,\text{m}\,\text{s}^{-1}$
and the acceleration of P is $(x^2 + 4)\,\text{m}\,\text{s}^{-2}$.
(a) Use the information given to form a differential equation
in the variables v and x only for the motion of P.
Given that $v = 2$ when $x = 0$,
(b) show that $3v^2 = 2x^3 + 24x + 12$.
(c) Find, in J, the work done on P by the force producing its
acceleration as P moves from $x = 0$ to $x = 9$.   [E]

16 A particle P, of mass m, moves along the x-axis under the action of a single force directed towards the origin O. The magnitude of the force is
$$\frac{mk}{2x + c}$$
where k and c are positive constants and x is the distance of P from O.

The particle is released from rest at the point A, where $OA = 2c$.

(a) Given that P is moving with speed v when it is at a distance x from O, where $0 \leqslant x \leqslant 2c$, show that
$$v^2 = k \ln\left(\frac{5c}{2x + c}\right)$$

(b) Hence, or otherwise, show that when P moves from B to O, where $OB = c$, the work done by the force is $\frac{1}{2}mk \ln 3$.   [E]

17 A particle P moves in a straight line. At time t seconds, the acceleration of P is $e^{2t}\,\mathrm{m\,s^{-2}}$, $t \geqslant 0$. When $t = 0$, P is at rest. Show that the speed, $v\,\mathrm{m\,s^{-1}}$, of P at time t seconds is given by
$$v = \tfrac{1}{2}\left(e^{2t} - 1\right)$$ [E]

18 A particle P, of mass m, moves along the x-axis under the action of a force directed towards the origin O. The magnitude of the force is $mk\,x^2$, where $OP = x$ and k is a positive constant. Given that P is initially at rest at the point C, where $x = c$, find

(a) the speed of P when it first passes through O, giving your answer in terms of c and k

(b) the work done by the force in moving P from C to O, giving your answer in terms of m, c and k.   [E]

19 A particle P, of mass m, moves away from the origin O along the positive x-axis under the action of a single force directed towards O of magnitude $\dfrac{\lambda m}{x^2}$, where λ is a positive constant. At the point C, where $OC = c$, the speed of P is $\sqrt{\left(\dfrac{\lambda}{c}\right)}$.

(a) Given that P is moving with speed v at the instant when it is at a distance x from O, show that
$$v^2 = \lambda\left(\frac{2c - x}{xc}\right)$$

(b) Hence, or otherwise, find in terms of λ, m and c the work done when P moves from the point C to the point D, where $OD = \dfrac{3c}{2}$.   [E]

20 An elastic string of natural length l has modulus of elasticity λ. One end of the string is fixed and the string hangs in equilibrium with a particle of mass m attached to the lower end. Find the stretched length of the string. [E]

21 A spring whose natural length is 200 mm has a length of 220 mm when supporting in equilibrium a mass of 10 kg. Calculate its length when supporting in equilibrium a mass of 25 kg. [E]

22 A particle of mass M rests suspended from a peg by a light elastic string. When $M = 1.5$ kg the length of the string is 0.45 m. When $M = 2$ kg the length of the string is 0.5 m. Find the unstretched length of the string. [E]

23 State Hooke's law for an elastic string.
An elastic string of natural length 1.2 m hangs vertically from a point O and has a body of mass 3 kg attached at the other end. The string hangs in equilibrium with the body 1.5 m below O. Calculate the modulus of elasticity of the string.
A new position of equilibrium, with the string inclined at an angle of 60° to the vertical, is maintained by a force applied to the body in a direction at right angles to the string.
Calculate
(a) the force applied,
(b) the tension in the string,
(c) the length of the string in this new position. [E]

24 Two light elastic strings AB and CD, each with the same natural length c, have moduli of elasticity $9Mg$ and $12Mg$ respectively. The ends B and C are joined together, the end A is tied at a fixed point and a particle of mass M is attached to D. The particle hangs in equilibrium with the strings vertical. Find, in terms of c, the length of AD. [E]

25 A particle P, of mass 6 kg, rests on a rough horizontal table at a distance of 15 cm from a vertical rod which is fixed to the table. A ring Q is free to move on the vertical rod and is attached to P by a light elastic string of natural length 15 cm. The string is such that when a mass of 3 kg is suspended freely at its lower end the extension is 15 cm. The ring Q is slowly raised up the vertical rod and has reached a height of 20 cm above the table when P is just about to slide on the table.

Find

(a) the tension in the string,

(b) the normal reaction between P and the table.

Prove that the coefficient of friction between P and the table is $\frac{3}{11}$. [E]

26

Two points A and B are 1.56 m apart in a horizontal line (see diagram). A body of mass 1.3 kg is suspended from A and B by two light elastic strings AC and BC. In the equilibrium position AC is perpendicular to BC and $\tan ABC = \frac{5}{12}$. Calculate, by resolving horizontally or otherwise, the ratio of the tensions in the two strings, and find the magnitude of the tension in the string AC.

Given that the unstretched length of the string AC is 54 cm, calculate the value of the modulus of elasticity of this string. [E]

27 A light elastic string, of natural length l and modulus of elasticity $4mg$, has one end tied to a fixed point A. The string passes over a fixed smooth peg B and at the other end a particle P, of mass m, is attached. The particle hangs in equilibrium. The distance between A and B is l and AB is inclined at $60°$ to the vertical as shown in the diagram.

(a) Find, in terms of l, the length of the vertical portion BP of the string.

(b) Show that the magnitude of the force exerted by the string on the peg is $mg\sqrt{3}$. [E]

28

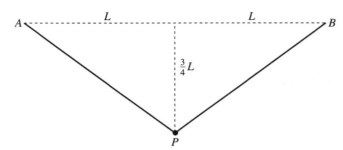

The diagram shows a particle P, of mass $6M$, suspended by two identical light elastic strings from the points A and B which are fixed and at a horizontal distance $2L$ apart. Each string has natural length L and P rests in equilibrium at a vertical distance $\frac{3}{4}L$ below the level of AB. Determine

(a) the tension in either string

(b) the modulus of elasticity of either string.   [E]

29 The diagram shows a particle C of mass $2\,\text{kg}$ suspended by two strings. The strings are fixed to points A and B on a horizontal ceiling, where $AB = 0.3\,\text{m}$. The string AC is light and inextensible, with length $0.4\,\text{m}$, while the string BC is light and elastic, with natural length $0.4\,\text{m}$ and modulus $32\,\text{N}$. A horizontal force of magnitude P newtons holds the system in equilibrium with AC vertical.

(a) Show that the tension in BC is $8\,\text{N}$.

(b) Find the value of P.

(c) Find, in N, the tension in AC.   [E]

30 A light elastic string of natural length a and modulus $7mg$ has a particle P of mass m attached to one end. The other end of the string is fixed to the base of a vertical wall. The particle P lies on a rough horizontal surface, and is released from rest at a distance $\frac{4}{3}a$ from the wall. The coefficient of friction between P and the surface is $\frac{1}{4}$.

Use the work–energy principle

(a) to show that P will hit the wall

(b) to find, in terms of a and g, the speed of P when it hits the wall.   [E]

31 A light elastic string of natural length 30 cm is placed on a smooth horizontal table with one end attached to a fixed point P on the table. The other end of the string is attached to a fixed point Q on the table such that $PQ = 80$ cm.

(a) Given that the tension in the string is 175 N, find, in newtons, the modulus of elasticity of the string.

The mid-point of the string is pulled a distance 30 cm along the perpendicular bisector of PQ in the plane of the table.

(b) Find the increase in the tension in the string.

(c) Find, in joules, the corresponding increase in the elastic energy of the string.   [E]

32 A particle P of mass $2m$ is fastened to one end of a light elastic string of natural length l and modulus of elasticity $4mg$. The other end of the string is fastened to a fixed point A. The particle hangs at rest at a point B vertically below A.

(a) Show that the extension of the string is $\dfrac{l}{2}$.

The particle is projected vertically downwards from B with speed $\sqrt{\left(\dfrac{gl}{2}\right)}$. When the particle is at a point C below B, where $BC = x$, its speed is v. Find:

(b) the increase in the elastic potential energy in the string as P moves from B to C, in terms of m, g, l and x

(c) the loss of kinetic energy plus the loss of gravitational potential energy of P as it moves from B to C, in terms of m, g, l, x and v.

Hence show that the particle first comes to instantaneous rest at the point D where $BD = \dfrac{l}{2}$.

(d) Find, in terms of g and l, the speed of P when it is at the mid-point of BD.   [E]

33 Show that the work done in stretching a light elastic string by a length x from its natural length a is $\frac{1}{2}Tx$, where T is the final tension in the string.

One end A of a light elastic string AB of natural length a is attached to a fixed point at a height $a + b$ above a horizontal floor, where $b > 0$. To the other end B is attached a small spherical ball of mass m. The modulus of elasticity of the string is $\frac{mga}{b}$.

(a) Show that in equilibrium under gravity the ball is just in contact with the floor.

The ball is released from rest at A.

(b) Show that it strikes the floor with speed v, given by
$$v^2 = (2a + b)g.$$

When the ball strikes the floor the elastic string breaks and the ball rebounds to a height a.

(c) Find, in terms of a and b, the coefficient of restitution between the ball and the floor. [E]

34 Using Hooke's law, show by integration that the work done in stretching a light elastic string, of natural length l and modulus of elasticity λ, from length l to length $l + e$ is $\frac{\lambda e^2}{2l}$.

A light elastic string, of natural length 2 metres and modulus of elasticity $4mg$ newtons, has one end attached to a fixed point A. A particle P, of mass m kilograms, is attached to the other end of the string.

(a) Show that, when P hangs in equilibrium, the length of AP is $2\frac{1}{2}$ m.

The particle P is released from rest at A and falls vertically.

Use the work–energy principle to calculate

(b) the speed, to 3 significant figures, of P at distance $2\frac{1}{2}$ m below A

(c) the greatest length of the string. [E]

35 A light elastic string AB has natural length l m and modulus of elasticity λ N. The end A is attached to a fixed point. When B is held at rest at a distance 0.6 m from A, the tension in the string is 3 N.

(a) Find an equation relating λ and l.

When B is held at rest at a distance 0.7 m from A, the elastic energy in the string is 0.6 J.

(b) Show that $l = 0.5$ and find the value of λ.

36 A light elastic string of natural length $30a$ is attached at its ends to points A and B which are in the same horizontal line and at a distance $30a$ apart. A particle of mass m is attached to the mid-point of the string and hangs in equilibrium at a distance $20a$ below AB.

(a) Show that the modulus of the string is $\frac{15}{16}mg$.

The particle is released from rest at O, the mid-point of the line AB. The particle has speed v when it is a distance x below O.

(b) Show that
$$v^2 = 2gx - \frac{1}{8}\left(\frac{g}{a}\right)\left[(225a^2 + x^2)^{\frac{1}{2}} - 15a\right]^2$$

(c) Hence, or otherwise, show that the particle never reaches a distance of $37a$ below O. [E]

37 A light elastic string of natural length $5\,\mathrm{m}$ and modulus $4\,\mathrm{N}$ has one end fixed to a point O on level ground. To the other end of the string is attached a ball of mass $0.5\,\mathrm{kg}$, which is projected vertically upwards from O with speed $u\,\mathrm{m\,s}^{-1}$.

(a) Find u, given that the ball first comes to instantaneous rest when it reaches a height of $10\,\mathrm{m}$ above the ground.

Given that the coefficient of restitution between the ball and the ground is 0.6

(b) show that, after the first bounce at O, the string does not become taut. [E]

38 A particle of mass m moves along the positive x-axis. It is acted upon by a force directed towards the origin and of magnitude mn^2x. Write down an equation governing the motion. Given that at time $t = 0$ the particle is at the origin and that the maximum distance of the particle from the origin is a, find x in terms of a, n and t. [E]

39 A particle is moving in a straight line so that at time t it is a distance s from a fixed point where
$$s = a\cos nt$$
a and n being constants.

Find:

(a) the velocity v and the acceleration f at time t

(b) an expression for $v^2 - fs$, and show that it is constant,

(c) an expression for $\dfrac{\mathrm{d}f}{\mathrm{d}t}$ and show that it is proportional to v.

Describe the motion of the particle along the straight line. [E]

40 A particle moves in a straight line with simple harmonic motion of period 4π seconds about a centre O. The speed of the particle is zero when it is at a distance 2 m from O. Find the speed when it is 1 m from O. **[E]**

41 A particle P starts from the point O at time $t = 0$ and moves in a straight line with velocity v, where

$$v = k\omega t \sin 2\omega t$$

k and ω being constant.

Find:

(a) the acceleration of P when $t = \pi/(2\omega)$

(b) the distance OP when P comes momentarily to rest for the first time after leaving O. **[E]**

42 A particle is moving along a straight line through a fixed point O and its displacement s metres, from O, at time t seconds, is given by the relation $s = 5\cos 2t$. Obtain expressions for the velocity and the acceleration of the particle at time t seconds and show how this acceleration is related to the displacement.

When $t = \dfrac{\pi}{6}$ the particle is at the point A and is moving towards O. Find:

(a) the distance OA

(b) the time when the particle next reaches O

(c) the velocity of the particle when it passes through O

(d) the acceleration of the particle when it reaches its furthest point from A

(e) the time, measured from $t = \dfrac{\pi}{6}$, that elapses before the particle next passes through A. **[E]**

43 A particle is moving along a straight line and its distance s metres from a fixed point O on the line at the instant t seconds after it was at O is $a\sin nt$ metres. Find the velocity and the acceleration of the particle t seconds after it was at O. The maximum speed of the particle is $0.6\,\mathrm{m\,s^{-1}}$ and its maximum acceleration is $2.4\,\mathrm{m\,s^{-2}}$. Find the values of a and n. Find also the time which elapses between the instant when the particle leaves O and when the particle is next at O. **[E]**

44 One end of a light elastic string, of natural length l, is attached to a fixed point O. To the other end of the string is attached a particle of mass m. When the particle hangs in equilibrium the length of the string is $3l/2$.

(a) The particle is pulled vertically downwards from its equilibrium position and released from rest when the length of the string is $7l/4$. Find the period and the amplitude of the subsequent oscillations.

(b) If instead the particle is released from rest at a point A vertically below O, where $OA = 9l/4$, show that the time taken for the string to first become slack is

$$\left(\frac{l}{2g}\right)^{\frac{1}{2}}\left[\pi - \arccos\left(\frac{2}{3}\right)\right]$$ [E]

45 (a) The level of the water in a harbour rises and falls vertically in simple harmonic motion. The depth of the water at low tide is 5.5 m and the depth at high tide is 11.5 m. A boat can enter the harbour when the depth of the water is 7 m. Low tide is at 0740 hours and high tide is at 1355 hours. Calculate the earliest time after 0740 hours at which the boat can just enter the harbour. [E]

(b) Describe briefly the model you used for the boat when investigating the motion in part (a) of this question.

46 A particle P of mass m hangs in equilibrium at one end of a vertical elastic string, the other end of which is attached to a fixed point. The extension of the string is e. If P is given a further extension, a, where $a < e$, and then released, find the period of its subsequent motion. [E]

47 A light elastic spring, of modulus $4mg$ and natural length l, has one end attached to a fixed point and carries a scale pan, of mass m, at the other end. Show that the period T of small oscillations of the pan about its equilibrium position is

$$2\pi\sqrt{(l/4g)}$$

When the pan is hanging in equilibrium, a particle of mass km is gently placed in the pan. Given that the periodic time of the ensuing motion is $2T$ find the value of k and the amplitude of the oscillation. [E]

48 A light elastic string of natural length l has one end A attached to a fixed point. A particle of mass m is attached to the other end B. When the particle hangs in equilibrium AB has length $\dfrac{5l}{4}$. The particle is pulled down vertically a further distance a and released from rest. Given that $a \leqslant \dfrac{l}{4}$ show that B first comes to instantaneous rest after a time $\frac{1}{2}\pi\sqrt{\left(\dfrac{l}{g}\right)}$ and find its greatest speed during this motion.

49 A light elastic string has natural length l and modulus $5mg$. One end of the string is attached to a fixed point A. A particle P of mass m hangs freely in equilibrium attached to the other end of the string.

(a) Show that P is a distance $\dfrac{6l}{5}$ vertically below A.

P is pulled down a vertical distance a below its equilibrium position and released from rest at time $t = 0$.

(b) Show that provided the string remains taut

$$\frac{\mathrm{d}^2 x}{\mathrm{d}t^2} = -\frac{5gx}{l}$$

where x is the displacement of P from its equilibrium position.

(c) Given that $a = \dfrac{l}{10}$ state the amplitude and find, in terms of l and g, the period of the resulting oscillations.

50 A particle P moves along a straight line so that, at time t seconds, its displacement x metres from a fixed point O on the line is given by

$$x = 3\cos\left(\tfrac{1}{2}\pi t\right)$$

(a) Find the acceleration of P at time t seconds and deduce that P is moving with simple harmonic motion.

(b) Write down the period of the motion of P.

(c) Write down, in terms of π, the maximum speed, V, of P.

(d) Find the first time, after $t = 0$, at which the speed of P is $\frac{1}{2}V$. [E]

51 A light elastic string, of natural length L and modulus $2mg$, has one end attached to a fixed point A. A particle P, of mass m, is attached to the other end of the string and hangs freely in equilibrium at the point O, vertically below A.

(a) Find, in terms of L, the length OA.

The particle P is pulled down a vertical distance h below O, and released from rest at time $t = 0$. At time t, the displacement of P from O is x.

(b) Show that, while the string is taut,
$$\frac{\mathrm{d}^2 x}{\mathrm{d}t^2} = -\frac{2gx}{L}$$

(c) State the set of values of h for which P performs complete simple harmonic oscillations.

Given that $h = \frac{1}{3}L$,

(d) find the time at which P first comes instantaneously to rest

(e) find the greatest speed of P.   [E]

Circular motion

4

When a particle P is moving on a circular path with centre O you can specify its position by measuring the angle between the radius OP and a fixed radius OA.

In the study of circular motion angles are measured in radians. Radians are discussed in Book P1, chapter 2.

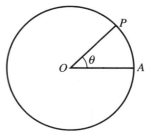

4.1 Angular speed

As P moves the angle θ changes. The rate at which θ changes with time is the **angular speed** ω of the radius OP. Since the angle θ is measured in radians the usual units for angular speed are radians per second ($\text{rad}\,\text{s}^{-1}$).

When it is necessary to distinguish between clockwise and anti-clockwise rotation the usual convention for angle measurement is used – that is, anticlockwise rotation is positive and clockwise rotation is negative.

The angular speed of the radius OP may also be described by the number of complete revolutions made in a given time interval. If the time interval is one minute the angular speed is measured in revolutions per minute ($\text{rev}\,\text{min}^{-1}$).

As ω is the rate of change of θ with respect to time it follows that:

■
$$\omega = \frac{\mathrm{d}\theta}{\mathrm{d}t}$$

When the angular speed ω is *constant* the particle is said to be moving in **uniform circular motion**.

It is important to remember that the angular speed ω refers to the speed at which the radius OP is turning. However, it is quite common to refer to the angular speed ω of the particle P about O even though strictly speaking this is not correct.

Example 1

A particle P is moving on a circular path centre O so that OP has a constant angular speed of $5\,\text{rad}\,\text{s}^{-1}$. Find the angular speed of OP in $\text{rev}\,\text{min}^{-1}$.

As: $$2\pi\ \text{rad} = 1\ \text{rev}$$

So: $$5\,\text{rad}\,\text{s}^{-1} = \frac{5}{2\pi}\ \text{rev}\,\text{s}^{-1}$$

And: $$1\ \text{minute} = 60\ \text{seconds}$$

So: $$\frac{5}{2\pi}\ \text{rev}\,\text{s}^{-1} = \frac{5}{2\pi} \times 60\ \text{rev}\,\text{min}^{-1}$$

$$= \frac{150}{\pi}\ \text{rev}\,\text{min}^{-1}$$

The angular speed of OP is $\dfrac{150}{\pi}\ \text{rev}\,\text{min}^{-1}$.

Example 2

A particle P is moving on a circular path centre O and is seen to make 10 complete revolutions in 4 minutes. Find the angular speed of P about O.

10 revolutions in 4 minutes is equivalent to $2.5\,\text{rev}\,\text{min}^{-1}$. As above, $1\,\text{rev} = 2\pi\,\text{rad}$ and $1\,\text{min} = 60\,\text{s}$

So: $$2.5\ \text{rev}\,\text{min}^{-1} = \frac{2.5 \times 2\pi}{60}\ \text{rad}\,\text{s}^{-1}$$

$$= \frac{\pi}{12}\ \text{rad}\,\text{s}^{-1}$$

The angular speed of P about O is $\dfrac{\pi}{12}\ \text{rad}\,\text{s}^{-1}$.

Angular and linear speed

Consider a particle P moving on a circular path centre O of radius r such that OP has constant angular speed ω. Take $\mathbf{i}$ and $\mathbf{j}$ to be unit vectors along and perpendicular to OA, the fixed direction from which θ is measured, as shown:

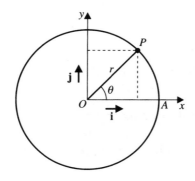

Suppose OP makes an angle θ with OA at time t seconds.

Since: $\qquad \dfrac{d\theta}{dt} = \omega \qquad$ and ω is constant,

integrating with respect to t gives:

$$\theta = \int \omega \, dt$$
$$= \omega t + c$$

where c is an arbitrary constant.

Taking $\theta = 0$ when $t = 0$ gives $c = 0$ and hence:

$$\theta = \omega t$$

That is: $\qquad\qquad\qquad \angle AOP = \omega t$

Hence when OP has constant angular speed ω it turns through an angle ωt in time t seconds. Writing the position vector $\mathbf{r}$ of P relative to O in component form gives

$$\mathbf{r} = r \cos \omega t \mathbf{i} + r \sin \omega t \mathbf{j}$$

as $\theta = \omega t$.

In Book M2, chapter 1 it was shown how vectors written in component form can be differentiated. In Book P3, chapter 2 it was shown that:

$$\frac{d}{dt}(\cos \omega t) = -\omega \sin \omega t$$

and: $\qquad\qquad \dfrac{d}{dt}(\sin \omega t) = \omega \cos \omega t$

So differentiating $\mathbf{r}$ with respect to time gives:

$$\mathbf{v} = \dot{\mathbf{r}} = -r\omega \sin \omega t \, \mathbf{i} + r\omega \cos \omega t \, \mathbf{j}$$

The linear speed v of the particle is the modulus of $\mathbf{v}$.

Thus: $\qquad\qquad v = |\mathbf{v}| = |-r\omega \sin \omega t \, \mathbf{i} + r\omega \cos \omega t \, \mathbf{j}|$

$$= r\omega |-\sin \omega t \, \mathbf{i} + \cos \omega t \, \mathbf{j}|$$

$$= r\omega \sqrt{\{\sin^2 \omega t + \cos^2 \omega t\}}$$

But $\sin^2 \omega t + \cos^2 \omega t = 1$ as shown in Book P1, chapter 2 so that v is constant and

∎ $\qquad\qquad\qquad\qquad v = r\omega$

When the radius is measured in metres and the angular speed in radians per second, the linear speed is in metres per second.

Direction of the linear velocity

If the particle P moving on a circle of radius r and centre O has position vector **r** relative to O given by

$$\mathbf{r} = r\cos\omega t\,\mathbf{i} + r\sin\omega t\,\mathbf{j}$$

then the velocity vector **v** of P is given by

$$\mathbf{v} = -r\omega\sin\omega t\,\mathbf{i} + r\omega\cos\omega t\,\mathbf{j}$$

as shown above.

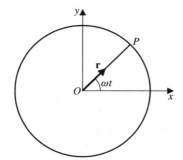

The gradient of the line OP is $\dfrac{r\sin\omega t}{r\cos\omega t} = \dfrac{\sin\omega t}{\cos\omega t}$

From the equation for **v** the gradient of the line representing the velocity vector **v** is

$$-\frac{r\omega\cos\omega t}{r\omega\sin\omega t} = -\frac{\cos\omega t}{\sin\omega t}$$

In Book P1, chapter 3 it was shown that:

if two lines have gradients m_1 and m_2 and

$$m_1 \times m_2 = -1$$

then the lines are perpendicular.

As : $\qquad \dfrac{\sin\omega t}{\cos\omega t} \times \left(-\dfrac{\cos\omega t}{\sin\omega t}\right) = -1$

it follows that the vectors $\overrightarrow{OP}$ and **v** are perpendicular. That is, the position vector **r** and the velocity vector **v** of P are perpendicular. Hence **the velocity is directed along the tangent to the circular path of P** as shown in the diagram below.

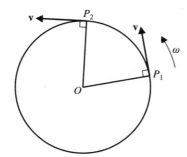

Angular speed is shown on diagrams with a curved arrow.

Example 3

A particle is moving on a circular path of radius 0.2 m, centre O, with constant angular speed $30 \, \text{rad s}^{-1}$ about O. Find its linear speed.

Using:
$$v = r\omega$$

with $r = 0.2$ and $\omega = 30 \, \text{rad s}^{-1}$ gives:
$$v = 0.2 \times 30$$
$$= 6$$

The linear speed of the particle is $6 \, \text{m s}^{-1}$.

Example 4

A particle is moving on a circular path of radius 0.75 m, centre O, with constant speed $5 \, \text{m s}^{-1}$. Find its angular speed about O in rev min^{-1}.

Using:
$$v = r\omega$$

with $r = 0.75$ m and $v = 5 \, \text{m s}^{-1}$ gives:
$$5 = 0.75\omega$$
$$\omega = \frac{5}{0.75}$$
$$\omega = \frac{20}{3}$$

The particle's angular speed about O is $\frac{20}{3} \text{rad s}^{-1}$.

So:
$$\text{angular speed} = \frac{20}{3} \times \frac{60}{2\pi} = \frac{200}{\pi} \text{rev min}^{-1}.$$

The particle's angular speed about O is $\frac{200}{\pi} \text{rev min}^{-1}$.

Example 5

A particle is moving on a circular path of radius 0.5 m at a constant speed of $10 \, \text{m s}^{-1}$. Find the time taken by P to complete 20 revolutions.

Using:
$$v = r\omega$$

with $v = 10 \, \text{m s}^{-1}$ and $r = 0.5$ m gives:
$$10 = 0.5\omega$$
$$\omega = 20$$

The angular speed is $20 \, \text{rad s}^{-1}$.

Since:
$$20 \text{ revolutions} = 20 \times 2\pi = 40\pi \text{ radians}$$

Using:
$$\theta = \omega t$$

with $\theta = 40\pi$ rad and $\omega = 20 \, \text{rad s}^{-1}$ gives:
$$40\pi = 20t$$
$$t = 2\pi = 6.28$$

The time taken is 6.28 seconds.

Exercise 4A

1   A particle is moving on a circular path with a constant angular speed of $3\,\text{rad}\,\text{s}^{-1}$. Find its angular speed in $\text{rev}\,\text{min}^{-1}$.

2   A particle is moving on a circular path with a constant angular speed of $4\,\text{rev}\,\text{min}^{-1}$. Find its angular speed in $\text{rad}\,\text{s}^{-1}$.

3   A particle moving on a circular path with constant speed is seen to complete 20 revolutions in 2 minutes. Find its constant angular speed (a) in $\text{rev}\,\text{min}^{-1}$ (b) in $\text{rad}\,\text{s}^{-1}$.

4   A particle moving on a circular path with constant speed completes 5 revolutions in 40 seconds. Find its constant angular speed (a) in $\text{rev}\,\text{min}^{-1}$ (b) in $\text{rad}\,\text{s}^{-1}$.

5   A particle is moving on a circular path of radius $0.5\,\text{m}$ with constant angular speed $25\,\text{rad}\,\text{s}^{-1}$. Find its constant linear speed.

6   A particle is moving on a circular path of diameter $0.5\,\text{m}$ with constant angular speed $3\,\text{rev}\,\text{min}^{-1}$. Find its constant linear speed.

7   The blade of a circular saw of radius $10\,\text{cm}$ is rotating at $4000\,\text{rev}\,\text{min}^{-1}$. Find the speed of a point on its rim in $\text{m}\,\text{s}^{-1}$.

8   A particle is moving on a circular path of radius $0.3\,\text{m}$ with a speed of $9\,\text{m}\,\text{s}^{-1}$. Find its angular speed in $\text{rad}\,\text{s}^{-1}$ and the time it takes to complete 6 revolutions.

9   A river which is flowing at $3.6\,\text{m}\,\text{s}^{-1}$ is turning a water wheel of radius $2\,\text{m}$. Assuming that the points on the rim of the wheel are moving at the speed of the water, calculate the angular speed of the wheel.

10   A car takes a bend by travelling along an arc of a circle of radius $35\,\text{m}$ at $14\,\text{m}\,\text{s}^{-1}$. Calculate the constant angular speed of the car.

11   Anne and John are standing on a moving roundabout $75\,\text{cm}$ and $125\,\text{cm}$ respectively from the centre. John's speed is $2\,\text{m}\,\text{s}^{-1}$. Find (a) the angular speed of the roundabout, assumed constant, (b) Anne's speed.

4.2 Acceleration in circular motion

When a particle is moving on a circular path centre O at a constant speed its direction of motion is continually changing. So the velocity of the particle is changing and the particle therefore has an acceleration.

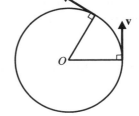

If a particle P is moving on a circle centre O of radius r, at time t seconds it has position vector $\mathbf{r}$ relative to O given by

$$\mathbf{r} = r\cos\omega t\,\mathbf{i} + r\sin\omega t\,\mathbf{j}$$

as shown in section 4.1.

Differentiating with respect to t gives:

$$\mathbf{v} = \dot{\mathbf{r}} = -r\omega\sin\omega t\,\mathbf{i} + r\omega\cos\omega t\,\mathbf{j}$$

Differentiating again with respect to t gives:

$$\mathbf{a} = \dot{\mathbf{v}} = \ddot{\mathbf{r}} = -r\omega^2\cos\omega t\,\mathbf{i} - r\omega^2\sin\omega t\,\mathbf{j}$$

$$\mathbf{a} = -\omega^2(r\cos\omega t\,\mathbf{i} + r\sin\omega t\,\mathbf{j})$$

But: $$\mathbf{r} = r\cos\omega t\,\mathbf{i} + r\sin\omega t\,\mathbf{j}$$

So: $$\mathbf{a} = -\omega^2\mathbf{r}$$

and the magnitude of $\mathbf{a}$, $|\mathbf{a}| = \omega^2 \times$ magnitude of $\mathbf{r}$

$$a = r\omega^2$$

Because of the minus sign, the direction of $\mathbf{a}$ is opposite to the direction of $\mathbf{r}$.

That is, the acceleration of P has magnitude $r\omega^2$ and is directed towards the centre of the circle.

As: $$v = r\omega$$

It follows that: $$\omega = \frac{v}{r}$$

and substituting for ω in $a = r\omega^2$ gives:

$$a = \frac{v^2}{r}$$

■ **So the acceleration of a particle moving on a circular path centre O of radius r is given by:**

$$a = r\omega^2 \quad \text{or} \quad a = \frac{v^2}{r}$$

and is directed towards the centre O of the circle.

Example 6

A particle is moving on a horizontal circular path of radius 0.5 m with a constant angular speed of 4 rad s^{-1}. Calculate its acceleration.

Using:
$$a = r\omega^2$$

with $r = 0.5$ m and $\omega = 4$ rad s^{-1} gives:
$$a = 0.5 \times 4^2$$
$$a = 8$$

The acceleration of the particle is 8 m s^{-2} towards the centre of the circle.

Exercise 4B

1 A particle is moving on a horizontal circular path of radius 0.75 m with a constant angular speed of 5 rev min^{-1}. Calculate the acceleration in m s^{-2} of the particle.

2 A particle is moving on a horizontal circular path of radius 2 m with a constant speed of 20 m s^{-1}. Determine the acceleration of the particle.

3 A particle is moving on a horizontal circular path of radius 1.5 m. The acceleration of the particle is 150 m s^{-2} towards the centre of the circle. Calculate (a) the angular speed, (b) the linear speed of the particle.

4 A car takes a bend by travelling along an arc of a circle of radius 35 m at 14 m s^{-1}. Calculate the acceleration of the car.

5 An insect flying around a light in a horizontal circle of radius 0.15 m is seen to make 30 revolutions per minute. Determine its acceleration.

6 Assuming the earth to be a sphere of radius 6370 km which makes one complete revolution in 24 hours, calculate the acceleration in m s^{-2} of a point on the equator.

4.3 Uniform motion of a particle moving in a horizontal circle

A particle which moves in a circle with constant speed has an acceleration directed towards the centre of the circle. There must be a force acting on the particle to produce this acceleration.

Using the equation of motion $F = ma$, along the radius, for a particle of mass m moving in a circle of radius r with constant speed v gives:

■
$$F = m\frac{v^2}{r} \quad \text{as} \quad a = \frac{v^2}{r}$$

So there is a force of magnitude $m\dfrac{v^2}{r}$ directed towards the centre of

the circle. This force can arise in several ways.

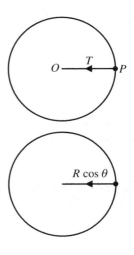

(1) A particle which is attached to one end of a string whose other end is fixed and moves in a horizontal circle will experience a force due to the tension in the string.

(2) A bead which is threaded on a smooth circular wire and is made to move round the wire will experience a force due to the reaction of the wire. The component of this force will act along the inward radius, thus providing the required force towards the centre.

(3) A car travelling round a bend, which is an arc of a circle, on a horizontal road will experience a frictional force between the road and its tyres.
    This force will be directed towards the centre O of the circular path and will produce the required acceleration if it is sufficiently large. However, there is a maximum frictional force that can result from the contact of two surfaces, as seen in Book M1, chapter 4. If this maximum frictional force is not sufficient to produce the required acceleration, the car will slip as it attempts to round the bend.

The force directed towards the centre of the circle is often referred to as the **centripetal force**.

Example 7
A particle P of mass $0.5\,\text{kg}$ is attached to one end of a light inextensible string of length $0.25\,\text{m}$. The other end of the string is fixed to a point O of a smooth horizontal table. P moves with speed $2\,\text{m s}^{-1}$ in a horizontal circle on the table. Calculate the tension in the string.

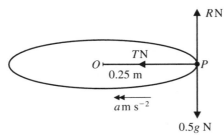

Let the tension in the string be $T\,\text{N}$ and the acceleration towards the centre be $a\,\text{m s}^{-2}$.

Using:
$$a = \frac{v^2}{r}$$

with $v = 2\,\text{m s}^{-1}$ and $r = 0.25\,\text{m}$ gives:

$$a = \frac{4}{0.25} = 16$$

The force required for the acceleration towards the centre is provided by the tension in the string.

Using:
$$F = ma$$

with $m = 0.5\,\text{kg}$ and $a = 16\,\text{m s}^{-2}$ gives:
$$T = 0.5 \times 16 = 8$$

The tension in the string is 8 N.

Example 8

A particle P of mass 0.5 kg rests on a rough horizontal disc at a distance 0.1 m from the centre. The disc is rotating about its centre. The coefficient of friction between the particle and the disc is $\frac{1}{4}$. Given that the particle is on the point of slipping, find the angular speed of the disc.

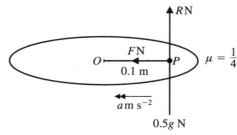

Here, the centripetal force is provided by the frictional force $F\text{N}$ between the particle and the disc. Let the (angular) speed of the disc be $\omega\,\text{rad s}^{-1}$ and the acceleration towards the centre be $a\,\text{m s}^{-2}$.

Using:
$$a = r\omega^2$$

with $r = 0.1\,\text{m}$ gives:
$$a = 0.1\omega^2$$

Using:
$$F = ma$$

with $m = 0.5\,\text{kg}$ and $a = 0.1\omega^2\,\text{m s}^{-2}$ gives:
$$F = 0.5 \times 0.1\omega^2$$
$$F = 0.05\omega^2$$

So the frictional force when the angular speed is $\omega\,\text{rad s}^{-1}$ is $0.05\omega^2$ N.

Since the particle is on the point of slipping:
$$F = \mu R$$

(from Book M1 chapter 4) and resolving vertically (no motion) gives:
$$R = 0.5g$$

As $\mu = \frac{1}{4}$ it follows that:
$$0.05\omega^2 = \frac{1}{4} \times 0.5g$$
$$\omega^2 = \frac{1}{4} \times \frac{0.5}{0.05}g$$

We require the speed, that is the magnitude of ω, and so we take the positive square root. (The sign of ω indicates the direction of its velocity.)

Hence: $$\omega = 4.949$$

The angular speed of the disc when the particle is on the point of slipping is $4.95\,\text{rad s}^{-1}$.

Example 9

A particle P of mass $0.25\,\text{kg}$ is attached to one end of a light elastic string of modulus $5\,\text{N}$ and natural length $0.75\,\text{m}$. The other end of the string is fixed to a point O on a smooth horizontal table. P moves in a horizontal circle of radius $1\,\text{m}$. Calculate the constant linear speed of P.

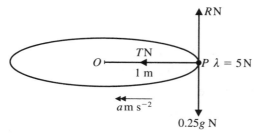

Let the speed of P be $v\,\text{m s}^{-1}$ and the tension in the string be $T\,\text{N}$. The tension can be obtained by using Hooke's law, which is explained in section 2.1.

Using Hooke's law: $$T = \frac{\lambda x}{l}$$

with $\lambda = 5\,\text{N}$, $l = 0.75\,\text{m}$ and $x = (1 - 0.75)\,\text{m} = 0.25\,\text{m}$ gives:

$$T = \frac{5 \times 0.25}{0.75}$$

$$T = \tfrac{5}{3}$$

The tension in the string is $\tfrac{5}{3}\,\text{N}$.

Using: $$F = ma$$

with $m = 0.25\,\text{kg}$ and $F = T\,\text{N} = \tfrac{5}{3}\,\text{N}$ gives:

$$\frac{5}{3} = 0.25a$$

$$a = \frac{5}{3 \times 0.25}$$

$$a = \frac{20}{3}$$

The acceleration of the particle is $\dfrac{20}{3}\,\text{m s}^{-2}$.

Using: $$a = \frac{v^2}{r}$$

with $a = \frac{20}{3}\,\mathrm{m\,s^{-2}}$ and $r = 1\,\mathrm{m}$ gives:

$$\frac{20}{3} = \frac{v^2}{1}$$

$$v^2 = \frac{20}{3}$$

and as $v \geqslant 0$: $\qquad\qquad v = 2.581$

The speed of P is $2.58\,\mathrm{m\,s^{-1}}$.

Exercise 4C

Whenever a numerical value of g is required take $g = 9.8\,\mathrm{m\,s^{-2}}$.

1 A particle P of mass 1 kg is attached to one end of a light inextensible string of length 0.3 m. The other end of the string is fixed to a point O of a smooth horizontal table. P moves with speed $5\,\mathrm{m\,s^{-1}}$ in a circle on the table. Calculate the tension in the string.

2 A particle P of mass 0.2 kg is attached to one end of a light inextensible string of length 0.5 m. The other end of the string is fixed to a point O of a smooth horizontal table. P describes horizontal circles, centre O, on the table. Given that the tension in the string is 2.5 N, calculate the speed of the particle.

3 A particle P of mass 0.2 kg rests on a rough horizontal disc at a distance of 0.3 m from the centre. The coefficient of friction between the particle and the disc is $\frac{1}{3}$. Given that the particle is on the point of slipping, calculate the angular speed of the disc.

4 A particle P of mass 0.5 kg rests on a rough horizontal disc at a distance of 0.25 m from the centre. When the disc is rotating with angular speed $5\,\mathrm{rad\,s^{-1}}$ P is on the point of slipping. Calculate the value of the coefficient of friction between P and the disc.

5 A smooth wooden hoop of radius 1 m rests on a smooth horizontal surface. A bead of mass 0.2 kg is moving in a horizontal circle in contact with the inner surface of the hoop. The bead maintains a constant speed of $6\,\mathrm{m\,s^{-1}}$. Calculate the magnitude of the reaction between the hoop and the bead.

6 A particle P of mass 2 kg is attached to one end of a light elastic string of modulus 20 N and natural length 0.5 m. The other end of the string is fixed to a point O on a smooth horizontal surface. P moves in a horizontal circle with centre O. Given that the speed of P is $10 \, \mathrm{m \, s}^{-1}$, calculate the radius of the circle.

7 A particle of mass 0.2 kg is attached to the end C of two light inextensible strings AC and BC both of length 1 m. The other ends A and B are attached to opposite ends of a diameter of a smooth disc of radius 0.8 m which is free to rotate in a horizontal plane about its centre. The disc rotates with angular speed $10 \, \mathrm{rad \, s}^{-1}$ and P remains in contact with the disc. (a) Calculate the tension in each string. (b) Given that the strings will break if the tension exceeds 30 N, calculate the maximum speed at which the disc can rotate without the strings breaking.

8 A particle of mass m is attached to one end of a light elastic string of natural length l and modulus $5 \, mg$. The other end of the string is fixed to a point O on a smooth horizontal table. The particle moves on the table in a circular path of radius $\dfrac{6l}{5}$ and centre O. Show that the constant speed u of the particle is given by

$$u = \sqrt{\frac{6gl}{5}}$$

9 OB is a light rigid rod of length $5l$. Particles of masses $3m$ and $4m$ are attached at points A and B of the rod, where OA is $2l$. The rod is made to rotate about O in a horizontal plane with constant angular speed ω. Calculate the ratio of the tensions in OA and AB.

10 A light rigid rod AB of length $6l$ has particles of masses $2m$ and $4m$ attached at A and B respectively. The rod rotates with constant angular speed ω in a horizontal plane about point D of the rod. The tensions in AD and DB are equal. Show that

$$AD = 4l$$

11 A bead P of mass 0.8 kg is attached to one end of a light inextensible string and rests on a smooth horizontal table. The

string passes through a small smooth hole, O, in the table and carries a mass of 0.5 kg which hangs freely. P is moving on the table in a circle of radius 0.4 m and centre O. Calculate the angular speed of P.

12 A bead B of mass 0.8 kg is threaded on a smooth circular wire centre O and radius 0.4 m which is fixed in a horizontal plane. A light inextensible string is attached to B and passes through a small smooth horizontal ring at O and carries a mass of 0.4 kg which is hanging freely. B is moving on the wire with a constant speed of 5 m s^{-1}. Calculate the magnitude and direction of the reaction of the wire on B.

The conical pendulum

A particle may move in a horizontal circle attached to a string whose other end is fixed not at the centre O of the circle but at a point A vertically above the centre. This situation is called a 'conical pendulum' since the string describes the surface of a cone as the particle moves around the circle.

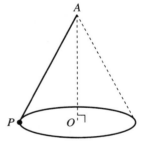

Example 10

A particle of mass 2 kg is attached to one end B of a light inelastic string AB of length 0.5 m. The other end A of the string is fixed. The particle moves with constant angular speed in a horizontal circle of radius 0.3 m whose centre O is vertically below A. Calculate the tension in the string and the angular speed of the particle.

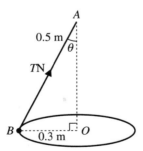

Let the tension in the string be T N and the angular velocity be ω rad s^{-1}.

The motion is taking place in a horizontal plane and the particle is not moving vertically.

So, resolving vertically:

$$T\cos\theta = 2g$$

From the diagram you can see that triangle OAB is a (3,4,5) triangle

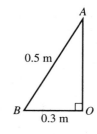

so: $$OA = 0.4\,\text{m}$$

and: $$\cos\theta = \tfrac{4}{5}$$

Hence: $$T \times \tfrac{4}{5} = 2g$$

$$T = \frac{10}{4}\,g = 24.5$$

The tension in the string is 24.5 N.

The particle is moving in a horizontal circle. Using the equation of motion:

$$F = ma \qquad \text{along } BO$$

gives: $$T\sin\theta = 2r\omega^2$$

Again from triangle OAB:

$$\sin\theta = \tfrac{3}{5}$$

and as $T = 24.5\,\text{N}$ and $r = 0.3\,\text{m}$ this gives:

$$24.5 \times \tfrac{3}{5} = 2 \times 0.3\omega^2$$

$$\omega^2 = \frac{24.5 \times 3}{5 \times 2 \times 0.3}$$

$$\omega = 4.949$$

as speed must be positive.

The angular speed is $4.95\,\text{rad s}^{-1}$.

Example 11

A ring R of mass 0.1 kg is threaded on a light inelastic string AB of length 3.5 m. The ends A and B of the string are attached to two fixed points 2.5 m apart in a vertical line. The ring is moving in a horizontal circle with constant angular speed and $\angle ARB$ is 90°.

(a) Suggest a suitable model for the ring so that the tension in the string can be calculated.
(b) By considering the forces acting on the ring show that

$$\frac{AR}{RB} > 1$$

(c) Prove that $AR = 2\,\text{m}$ and find the radius of the circular path described by the ring.
(d) Calculate the tension in the string and the angular speed of the ring.

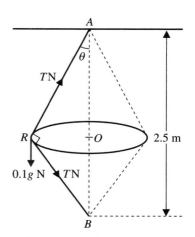

As the ring is only *threaded* on the string and there is no knot at the ring the string is continuous. Consequently the tensions in AR and BR are the same. Let these tensions be T N and let $\angle RAB$ be θ.

(a) The ring can be modelled as a particle as it is small compared with the length of the string.

(b) As there is no vertical motion, resolving vertically gives:

$$T\cos\theta = 0.1g + T\sin\theta \tag{1}$$

So:
$$T(\cos\theta - \sin\theta) = 0.1g$$

As $T > 0$ it follows that:

$$\cos\theta - \sin\theta > 0$$

and so:
$$\cos\theta > \sin\theta$$

As:

$$\frac{\sin\theta}{\cos\theta} = \tan\theta$$

(Book P1, chapter 2) it follows that:

$$\tan\theta < 1$$

From triangle ARB, which is right-angled at R:

$$\tan\theta = \frac{BR}{AR}$$

And so:
$$\frac{BR}{AR} < 1$$

or:
$$\frac{AR}{BR} > 1$$

(c) Let AR be x m.

Then $BR = (3.5 - x)$ m as the string is 3.5 m long. As triangle ARB is right-angled, using Pythagoras' theorem gives:

$$AR^2 + BR^2 = AB^2$$

$$x^2 + (3.5 - x)^2 = 2.5^2$$

$$x^2 + 12.25 - 7x + x^2 = 6.25$$

$$2x^2 - 7x + 6 = 0$$

$$(2x - 3)(x - 2) = 0$$

$$x = \tfrac{3}{2} \quad \text{or} \quad x = 2$$

The two solutions for x are the lengths of AR and BR as their sum is 3.5.

From part (b):
$$AR > BR$$

So:
$$AR = 2m$$

To calculate the radius of the circle described by R consider triangle AOR.

Let the radius be r m.

Then:
$$r = AR\sin\theta$$

From triangle ARB:
$$\sin\theta = \frac{RB}{AB} = \frac{1.5}{2.5} = \frac{3}{5} \qquad (2)$$

So:
$$r = 2 \times \tfrac{3}{5} = \tfrac{6}{5}$$

The radius of the circular path is $1\frac{1}{5}$ m.

(d) Equation (1) gives:
$$T\cos\theta = 0.1g + T\sin\theta$$

Equation (2) gives:
$$\sin\theta = \tfrac{3}{5}$$

and hence:
$$\cos\theta = \tfrac{4}{5}$$

So:
$$T \times \tfrac{4}{5} - T \times \tfrac{3}{5} = 0.1g$$
$$\tfrac{1}{5}T = 0.1 \times 9.8$$
$$T = 4.9$$

The tension is 4.9 N.

To find the angular speed consider the equation of motion along the inward radius.

Using:
$$F = ma$$

with $m = 0.1$ kg gives:
$$T\sin\theta + T\cos\theta = 0.1r\omega^2$$

Substituting $T = 4.9$, $\sin\theta = \tfrac{3}{5}$, $\cos\theta = \tfrac{4}{5}$ and $r = \tfrac{6}{5}$ gives:

$$4.9 \times \left(\tfrac{3}{5} + \tfrac{4}{5}\right) = 0.1 \times \tfrac{6}{5} \times \omega^2$$
$$\omega^2 = \frac{4.9 \times 7}{0.1 \times 6}$$
$$\omega = 7.560 \quad \text{(as speed must be positive)}$$

The angular speed is 7.56 rad s^{-1}.

Example 12

A child's toy consists of a small sphere P of mass m joined to two rods AP, BP each of length l.

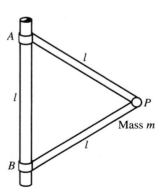

The other ends of the rods are a distance l apart on a third rod as shown in the diagram. AP and BP can rotate about AB. Assuming that the rods AP and BP are light and AB is vertical, calculate the tensions in the rods AP and BP when P is moving in a circle with speed $\sqrt{(3gl)}$. State clearly the model you have used for P in your solution and any other assumptions made.

As P is a small sphere it can be modelled as a particle. It is also assumed that the rods AP and BP rotate freely about AB, that is there are no frictional forces at A and B.

Let the tensions in AP and BP be T_1 and T_2 respectively. Let P move in a circle centre O, the mid-point of AB. Triangle APB is equilateral as all sides have length l. Hence $\angle A = 60°$ and $OP = l\cos 30°$.

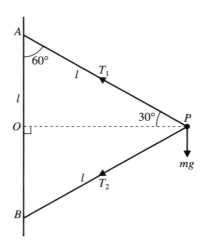

Using the equation of motion, $F = ma$, along the radius OP gives:

$$T_1 \cos 30° + T_2 \cos 30° = m\,\frac{v^2}{r}$$

But $r = l\cos 30°$ from above and $v = \sqrt{(3gl)}$

So:
$$T_1 \cos 30° + T_2 \cos 30° = m \times \frac{3gl}{l\cos 30°}$$

$$T_1 \cos^2 30° + T_2 \cos^2 30° = 3mg$$

But $\cos 30° = \dfrac{\sqrt{3}}{2}$ and so $\cos^2 30° = \tfrac{3}{4}$

Hence:
$$\frac{3T_1}{4} + \frac{3T_2}{4} = 3mg$$

and so:
$$T_1 + T_2 = 4mg \qquad\qquad (1)$$

There is no vertical motion, so resolving vertically gives:

$$T_1 \cos 60° = T_2 \cos 60° + mg$$
$$\tfrac{1}{2}T_1 = \tfrac{1}{2}T_2 + mg$$
$$T_1 - T_2 = 2mg \qquad\qquad (2)$$

Adding equations (1) and (2) gives:

$$2T_1 = 6mg$$
$$T_1 = 3mg$$

And substituting this value of T_1 in either equation gives:

$$T_2 = mg$$

The tensions in AP and BP are $3mg$ and mg respectively.

Motion on a banked surface

When a car is travelling round a circular bend it will slip if the frictional force between its tyres and the road is insufficient to produce the required acceleration towards the centre of the circle.

However the magnitude of the force towards the centre of the circular path can be increased if the road surface is inclined to the horizontal. Such a surface is said to be **banked** and it can be seen from the diagram that the normal reaction and the frictional force now both have components towards the centre of the circle. This allows the car to proceed at a higher speed than would otherwise have been possible.

If an angle between the road surface and horizontal is α the road is said to be *banked at an angle α to the horizontal.*

Example 13

A boy is cycling at a constant speed of $10 \, \text{m s}^{-1}$ round a circular bend of radius 30 m. The boy and his bicycle can be modelled as a single particle of mass 50 kg. He can be assumed to be moving in a horizontal circle and the road is assumed to be smooth. At what angle to the horizontal is the road banked if no slipping occurs?

Let the path be banked at an angle α to the horizontal as shown in the diagram.

The path is smooth so the only forces acting on the boy and his cycle are their combined weight and the normal reaction.

The acceleration towards the centre $= \dfrac{v^2}{r} = \dfrac{10^2}{30} = \dfrac{10}{3} \, \text{m s}^{-2}$

Using: $\qquad\qquad\qquad F = ma$

horizontally gives: $\qquad R \sin \alpha = 50 \times \dfrac{10}{3}$ $\qquad\qquad$ (1)

There is no vertical motion, so:

$$R \cos \alpha = 50g \qquad\qquad (2)$$

Eliminating R between equations (1) and (2) gives:

$$\frac{\sin \alpha}{\cos \alpha} = \frac{10 \times 50}{3} \div 50g$$

$$\tan \alpha = \frac{10}{3g} = \frac{10}{3 \times 9.8}$$

So: $\qquad\qquad\qquad \alpha = 18.8°$

The road should be banked at $18.8°$ to the horizontal.

Example 14
A car of mass 1500 kg is rounding a bend on a road which is banked at $15°$ to the horizontal. The car is assumed to be moving at a constant speed in a horizontal circle of radius 300 m. The coefficient of friction between the car's tyres and the road is 0.3. Calculate the maximum speed at which the car can travel if slipping is not to occur.

Let the maximum possible speed by $v \, \text{m s}^{-1}$.

Then the acceleration towards the centre will be $\dfrac{v^2}{300} \, \text{m s}^{-2}$.

Using: $\qquad\qquad\qquad\qquad F = ma$

horizontally gives: $\quad R \sin 15° + F \cos 15° = 1500 \times \dfrac{v^2}{300}$

So: $\qquad\qquad R \sin 15° + F \cos 15° = 5v^2 \qquad\qquad (1)$

There is no vertical motion so:

$$R \cos 15° = F \sin 15° + 1500g \qquad\qquad (2)$$

At maximum speed, the car is on the point of slipping so friction is limiting.

That is: $\qquad\qquad\qquad F = \mu R = 0.3R$

Substituting in equation (1) gives:

$$R \sin 15° + 0.3R \cos 15° = 5v^2$$

$$R(\sin 15° + 0.3 \cos 15°) = 5v^2 \qquad\qquad (3)$$

and in equation (2) gives:

$$R\cos 15° = 0.3R\sin 15° + 1500g$$

$$R(\cos 15° - 0.3\sin 15°) = 1500g \qquad (4)$$

Dividing equations (3) and (4) gives:

$$\frac{\sin 15° + 0.3\cos 15°}{\cos 15° - 0.3\sin 15°} = \frac{5v^2}{1500g}$$

$$v^2 = 300g\left(\frac{\sin 15° + 0.3\cos 15°}{\cos 15° - 0.3\sin 15°}\right)$$

$$v = 42.6$$

The maximum speed is $42.6\,\mathrm{m\,s^{-1}}$.

When a car is travelling round a circular bend which is banked it can slip *down* the slope if the friction force acts up the slope and is not sufficient to prevent sliding. This will arise when the acceleration is such that the component of the normal reaction alone is greater than the force required to produce the acceleration.

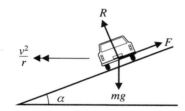

That is, $R\sin \alpha > \dfrac{mv^2}{r}$. Friction must therefore be acting up the plane and for motion without slipping using $F = ma$ horizontally gives:

$$R\sin \alpha - F\cos \alpha = \frac{mv^2}{r} \qquad (1)$$

However, as $F \leqslant \mu R$ sometimes it will not be possible to increase F sufficiently to satisfy equation (1) and the car will slip down the slope.

Example 15

A girl is cycling round a circular bend of radius 40 m. The road is banked at an angle of 20° to the horizontal and the coefficient of friction between the bicycle's tyres and the road is 0.3. The girl and her bicycle have a combined mass of mass 60 kg and can be assumed to be travelling in a horizontal circle. Find the least speed at which she can cycle without slipping.

Let the minimum possible speed be $v \, \mathrm{m \, s^{-1}}$.

Then the acceleration towards the centre will be $\dfrac{v^2}{40} \, \mathrm{m \, s^{-2}}$.

Using:
$$F = ma$$

horizontally gives:

$$R \sin 20° - F \cos 20° = 60 \times \frac{v^2}{40}$$

So:
$$R \sin 20° - F \cos 20° = \frac{3v^2}{2} \qquad (1)$$

There is no vertical motion so:
$$R \cos 20° + F \sin 20° = 60g \qquad (2)$$

At minimum speed, the bicycle is on the point of slipping so friction is limiting.

That is:
$$F = \mu R = 0.3R$$

Substituting in equation (1) gives:

$$R \sin 20° - 0.3R \cos 20° = \frac{3v^2}{2}$$

$$R(\sin 20° - 0.3 \cos 20°) = \frac{3v^2}{2} \qquad (3)$$

and in equation (2) gives:

$$R \cos 20° + 0.3R \sin 20° = 60g$$
$$R(\cos 20° + 0.3 \sin 20°) = 60g \qquad (4)$$

Dividing equations (3) and (4) gives:

$$\frac{\sin 20° - 0.3 \cos 20°}{\cos 20° + 0.3 \sin 20°} = \frac{3v^2}{120g}$$

$$v^2 = 40g \left(\frac{\sin 20° - 0.3 \cos 20°}{\cos 20° + 0.3 \sin 20°} \right)$$

$$v = 4.75$$

The minimum speed is $4.75 \, \mathrm{m \, s^{-1}}$.

Exercise 4D

1   A glass marble of mass m is moving in a horizontal circle round the inside surface of a smooth hemispherical bowl of radius r. The centre of the circle is at a distance $\frac{1}{2}r$ below the centre of the bowl. State a suitable model for the marble and hence find the magnitude of the reaction between the marble and the bowl and the speed of the marble.

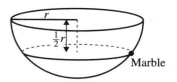

2   A boy sets his toy car running round the smooth sloping side of a circular sand-pit. The car moves in a horizontal circle of radius $0.75\,\text{m}$ and the sides of the sand-pit are inclined at an angle $\arcsin\frac{4}{5}$ to the horizontal. State suitable models for the car and the side of the sand-pit and hence calculate the speed of the car.

3   A particle of mass $0.8\,\text{kg}$ is attached to end B of a light inelastic string AB of length $0.6\,\text{m}$, the other end of which is fixed at A. The particle moves with a constant angular speed in a horizontal circle of radius $0.2\,\text{m}$ whose centre is vertically below A. Calculate the tension in the string and the angular speed of the particle.

4   A particle of mass $1.5\,\text{kg}$ is attached to end B of a light inelastic string AB of length $0.5\,\text{m}$, the other end of which is fixed at A. The particle moves in a horizontal circle whose centre is vertically below A with a constant angular speed. The particle takes $\frac{\pi}{4}\,\text{s}$ to complete one revolution. Calculate the tension in the string and the radius of the circular path.

5   A particle P of mass $2\,\text{kg}$ is attached to one end of a light inextensible string of length $1.2\,\text{m}$. The other end of the string is fixed at a point A which is $0.6\,\text{m}$ above a smooth horizontal table. P moves on the table on a circular path whose centre O is vertically below A. Given that the string is taut and the angular speed of P about A is $3\,\text{rad s}^{-1}$ calculate (a) the tension in the string (b) the normal reaction between the particle and the table.

6   The end A of a light inextensible string of length l is attached to a fixed point. A particle P of mass m is attached to the free end. P moves with constant speed $2v$ in a horizontal circle

with centre vertically below A and the string remains taut, inclined at an angle ϕ to the vertical. (a) Find v^2 in terms of l, g and ϕ. (b) Given also that the string breaks when the tension exceeds $4mg$, show that $v \leqslant \frac{1}{4}\sqrt{15lg}$.

7 A light elastic string AB has natural length l and modulus $\frac{20mg}{3}$. The end A is fixed and a particle of mass m is attached to the end B.

The particle is made to move in a horizontal circle at a constant angular speed with the string inclined at an angle $\arcsin\frac{4}{5}$ to the vertical. Find, in terms of l and g, the angular speed of the particle.

8 A light inextensible string AB of length $3l$ has its ends fixed to two points A and B which are in a vertical line with A a distance l above B. A smooth ring of mass m is threaded on the string and is made to move in a horizontal circle centre B with a constant speed. (a) State a suitable model for the ring and hence calculate the tension in the string and the speed of the ring. (b) Explain briefly the significance of the statement 'The ring is threaded on the string' in the solution of the problem.

9 A small smooth ring of mass m is threaded on a light inelastic string of length $9l$, whose ends are fixed at a point O on a smooth horizontal table and at a point A which is at a height $6l$ vertically above O. The ring moves in a horizontal circle on the table with the string taut. Given that the ring is moving at a constant angular speed $\frac{1}{2}\sqrt{\left(\frac{g}{l}\right)}$ calculate the magnitude of the reaction between the ring and the table.

10 A particle P is attached to the ends of two identical light inelastic strings which have their other ends fixed to two points A and B where A is $2l$ above B in a vertical line. P moves in a horizontal circle with constant angular speed $4\sqrt{\frac{g}{l}}$ and both strings are taut. Calculate the ratio of the tensions in AP and BP.

11 A particle P is attached to the ends of two elastic strings which have the same natural length l. The other ends of the strings are fixed to two points A and B where A is $3l$ above B in a vertical line. The modulus of the string BP is one third that of AP. P moves in a horizontal circle with centre at the mid-point of AB and both strings taut. Find the angular speed of P in terms of g and l.

12 A girl is cycling round a bend which forms an arc of a circle of radius 20 m. The track she is cycling on is banked at an angle of 25° to the horizontal. There is no frictional force acting on the cyclist. By modelling the girl and her cycle as a single particle of mass 70 kg, find the speed of the cyclist.

13 A car is travelling round a bend in a road which is banked at an angle of 30° to the horizontal. The bend is assumed to be in the shape of an arc of a circle of radius 80 m. The surface of the road is rough, the coefficient of friction between the tyres of the car and the surface of the road is 0.3. Find (a) the greatest speed and (b) the least speed at which the car can travel round the bend without slipping occurring.

14 A car is travelling round a circular bend of radius 70 m. In an initial model the road is assumed to be horizontal. The car is travelling at a constant speed of 16 m s^{-1} and no slipping occurs. (a) Calculate the least value of the coefficient of friction between the car's tyres and the road. In a revised model, the road is assumed to be banked at an angle of 15° to the horizontal. The speed of the car is such that it can travel round the bend with no friction force between its tyres and the road. (b) Calculate the speed at which the car is travelling.

4.4 The motion of a particle in a vertical circle

A particle can be caused to move on a path which is part of a circle in a vertical plane in a variety of ways. For example, it could be attached to the end of a string or rod and set in motion, or it could be placed on top of a cylinder whose axis is horizontal and then disturbed so that it slides. For the circular motion in a horizontal plane studied in section 4.3, the angular speed was constant. For motion in a vertical circle this is not the case, so the first step is to investigate the acceleration of the particle once more.

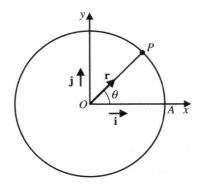

Let the particle P, which is constrained to move in a vertical plane on a circular path centre O of radius r, have position vector $\mathbf{r}$ relative to O.

Let the angle between $\mathbf{r}$ and OA, a fixed line through O, be θ at time t.

Taking unit vectors $\mathbf{i}$ and $\mathbf{j}$ along and perpendicular to OA gives:

$$\mathbf{r} = r\cos\theta\,\mathbf{i} + r\sin\theta\,\mathbf{j}$$

Differentiating with respect to *time*, as shown in Book P3, chapter 2, gives:

$$\mathbf{v} = \dot{\mathbf{r}} = -r\sin\theta\dot{\theta}\,\mathbf{i} + r\cos\theta\dot{\theta}\,\mathbf{j}$$

Differentiating again with respect to time, using the product rule for differentiation (Book P3, chapter 2):

$$\mathbf{a} = \ddot{\mathbf{r}} = (-r\cos\theta\dot{\theta}^2 - r\sin\theta\ddot{\theta})\mathbf{i} + (-r\sin\theta\dot{\theta}^2 + r\cos\theta\ddot{\theta})\mathbf{j}$$

This can be rearranged to give:

$$\ddot{\mathbf{r}} = -(r\cos\theta\,\mathbf{i} + r\sin\theta\,\mathbf{j})\dot{\theta}^2 + (r\sin\theta\,\mathbf{i} - r\cos\theta\,\mathbf{j})\ddot{\theta}$$
$$= -\mathbf{r}\dot{\theta}^2 + (r\sin\theta\,\mathbf{i} - r\cos\theta\,\mathbf{j})\ddot{\theta}$$

There are two separate parts to this acceleration:

(1) $-\mathbf{r}\dot{\theta}^2$ which is a vector of magnitude $r\dot{\theta}^2$ directed along PO (this is equal to $-\omega^2 r$ when $\dot{\theta}$ is constant, as for motion in a horizontal circle).

(2) $(r\sin\theta\,\mathbf{i} - r\cos\theta\,\mathbf{j})\ddot{\theta}$ (this is zero when $\dot{\theta} = \omega = $ constant).

Now:
$$|(r\sin\theta\,\mathbf{i} - r\cos\theta\,\mathbf{j})\ddot{\theta}| = |\sin\theta\,\mathbf{i} - \cos\theta\,\mathbf{j}|r\ddot{\theta}$$
$$= r\ddot{\theta}\sqrt{(\sin^2\theta + \cos^2\theta)}$$
$$= r\ddot{\theta} \quad \text{as } \sin^2\theta + \cos^2\theta = 1$$

So the magnitude of $(r\sin\theta\,\mathbf{i} - r\cos\theta\,\mathbf{j})\ddot{\theta}$ is $r\ddot{\theta}$.

To find its direction:

The gradient of the line representing $(r\sin\theta\,\mathbf{i} - r\cos\theta\,\mathbf{j})\ddot{\theta}$ is:

$$-\frac{r\cos\theta}{r\sin\theta} = -\frac{\cos\theta}{\sin\theta}$$

and the gradient of the line representing **r** is:

$$\frac{r\sin\theta}{r\cos\theta} = \frac{\sin\theta}{\cos\theta}$$

Since $\qquad \dfrac{\sin\theta}{\cos\theta} \times -\dfrac{\cos\theta}{\sin\theta} = -1$

these two lines are perpendicular as shown in Book P1, chapter 3.

So $r\sin\theta\,\mathbf{i} - r\cos\theta\,\mathbf{j}$ is perpendicular to **r** and is directed along the tangent to the circular path. Hence the acceleration has two components:

$r\dot{\theta}^2$ along the inward radius towards O

and $\quad r\ddot{\theta}\quad$ along the tangent to the circular path.

$\dot{\theta}$ is the rate of change of θ with respect to time, that is, $\dot{\theta}$ is the angular speed of OP.

Writing this angular speed as ω, but remembering that for motion in a vertical circle ω *is not constant* gives:

$$\dot{\theta} = \omega = \frac{v}{r}$$

where v is the linear speed of P.

And so: $\qquad\qquad\qquad r\dot{\theta} = v$

Differentiating with respect to time gives:

$$r\ddot{\theta} = \frac{dv}{dt} \quad \text{as } r \text{ is constant.}$$

The components of the acceleration are thus:

■ $\qquad r\omega^2 \ \text{ or } \ \dfrac{v^2}{r} \qquad$ **along the inward radius**

and $\quad \dfrac{dv}{dt}$ **along the tangent.**

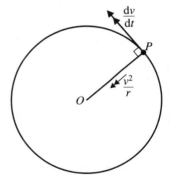

In some cases of motion in a vertical circle, the particle, once in motion, is subject only to gravity and a force along the radius of the circular path from, for example, the tension in the string or rod. As this radial force is perpendicular to the direction of motion of the particle it does no work. It follows that the sum of the kinetic energy and potential energy of the particle is constant throughout the motion. This is an application of the **work–energy principle** which states that:

■ **work done by the forces** $\quad = \quad$ **change in mechanical**
    acting on a particle $\qquad\qquad$ **energy of the particle**

This fact is very important in solving problems as it enables you to calculate the speed of the particle at a given point and hence also the radial acceleration and the force acting on the particle.

Motion of a particle on a fixed vertical circle

A particle which is attached to one end of a rod or is threaded on a circular wire and caused to move in a vertical circle cannot leave that circular path. In some cases the particle will describe complete circles but in other cases it will move on an arc of a circle.

Consider a particle P of mass m which is attached to one end of a light rod of length r. The rod is free to rotate in a vertical plane about an axis through the other end O. Suppose P is hanging in equilibrium and is then given a speed u.

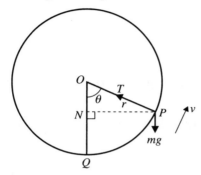

Let the speed of P be v when OP makes an angle θ with the downward vertical.

By the work–energy principle:

as the tension is perpendicular to the direction of motion and hence does no work,

$$\text{potential energy gained} = \text{kinetic energy lost}$$

The particle has risen a vertical distance $OQ - ON$.

As $ON = r\cos\theta$, the distance $OQ - ON$ is $r - r\cos\theta$.

So: P.E. gained $= mgh = mg(r - r\cos\theta)$

And: K.E. lost $= \frac{1}{2}mu^2 - \frac{1}{2}mv^2$

Hence: $mg(r - r\cos\theta) = \frac{1}{2}mu^2 - \frac{1}{2}mv^2$

Rearranging this gives:

$$\frac{1}{2}mv^2 = \frac{1}{2}mu^2 - mg(r - r\cos\theta)$$

And so: $v^2 = u^2 - 2gr(1 - \cos\theta)$ (1)

If P is to describe complete circles, it must have a speed at the highest point, that is $v^2 > 0$ when $\theta = 180°$.

When $\theta = 180°$: $v^2 = u^2 - 2gr(1 - \cos 180°)$

$$v^2 = u^2 - 2gr(1 - (-1))$$

$$v^2 = u^2 - 4gr$$

And for $v^2 > 0$:

$$u^2 - 4gr > 0$$
$$u^2 > 4gr$$

So a particle moving on a fixed vertical circle of radius r will move in complete circles if its speed u at the lowest point satisfies

$$u^2 > 4gr$$

If this is not the case, that is if $u^2 - 4gr < 0$, the speed must become zero before the particle reaches the highest point of the circle and the particle will oscillate on the arc ABC of the circle as shown in the diagram.

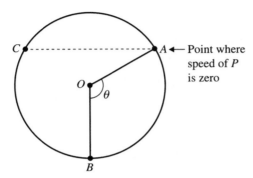

From equation (1) above, the speed v at any point on the path is given by:

$$v^2 = u^2 - 2gr(1 - \cos\theta)$$

and substituting $v = 0$ enables you to calculate the maximum value of θ:

$$0 = u^2 - 2gr(1 - \cos\theta_{max})$$
$$0 = u^2 - 2gr + 2gr\cos\theta_{max}$$
$$\cos\theta_{max} = \frac{2gr - u^2}{2gr}$$
$$\theta_{max} = \arccos\left(\frac{2gr - u^2}{2gr}\right)$$

Example 16

A particle of mass 0.5 kg is attached to one end A of a light rod AB of length 0.2 m which is free to rotate in a vertical plane about the other end B. The particle is held at rest vertically above B. It is slightly displaced so that it moves in a vertical circle.

Calculate:

(a) the speed of the particle as it passes through the lowest point of its path
(b) the tension in the rod at this point.

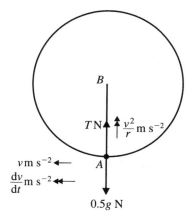

(a) Let the speed of the particle at its lowest point be $v\,\mathrm{m\,s}^{-1}$ and the tension in the rod be T N.

At the lowest point the particle has moved a distance of 0.4 m downwards from its initial position.

So: potential energy lost $= mgh = 0.5 \times 0.4\,\mathrm{J}$

and: kinetic energy gained $= \tfrac{1}{2}\,mv^2 - 0 = \tfrac{1}{2} \times 0.5v^2\,\mathrm{J}$

Using the work–energy principle:

$$\text{K.E. gained} = \text{P.E. lost}$$

So: $\tfrac{1}{2} \times 0.5 \times v^2 = 0.5 \times g \times 0.4$

$$v^2 = 9.8 \times 0.8$$

$$v = 2.8$$

The speed of the particle at its lowest point is $2.8\,\mathrm{m\,s}^{-1}$.

(b) To calculate the tension in the rod, write down the vertical component of the equation of motion.

Using: $F = ma$

gives: $T - 0.5g = 0.5\,\dfrac{v^2}{r}$

Since $v = 2.8\,\mathrm{m\,s}^{-1}$ and $r = 0.2\,\mathrm{m}$ this gives:

$$T = 0.5 \times 9.8 + 0.5 \times \frac{2.8^2}{0.2}$$

and so: $T = 24.5$

The tension in the rod is 24.5 N.

Example 17

A bead P is threaded on a smooth circular wire of radius 0.5 m and centre O which is fixed in a vertical plane. The bead is projected from the lowest point of the wire with a speed $u\,\mathrm{m\,s}^{-1}$.

(a) Describe the model you would use for the bead in order to investigate its motion.

(b) Show that if $u^2 > 19.6$ the bead will describe complete circles.

(c) Given that $u^2 = 4.9$ calculate the angle between OP and the downward vertical at the highest point of P's path.

(a) The wire has radius 0.5 m and relative to that the bead's radius is small, so the bead can be modelled as a particle.

(b) For the bead to describe complete circles, it must have a speed at the highest point of the circular path. Let this speed be $v\,\text{m s}^{-1}$. Then $v^2 > 0$ if the bead is to describe complete circles.

Using the work–energy principle:

$$\text{K.E. lost} = \text{P.E. gained}$$

$$\tfrac{1}{2}mu^2 - \tfrac{1}{2}mv^2 = mgh$$

where $m\,\text{kg}$ is the mass of the bead.

But: $\qquad\qquad\qquad h = 2 \times \text{radius} = 2 \times 0.5 = 1$

So: $\qquad\qquad\quad \tfrac{1}{2}mu^2 - \tfrac{1}{2}mv^2 = m \times 9.8 \times 1$

$$u^2 - v^2 = 9.8 \times 2$$

$$v^2 = u^2 - 19.6$$

Since: $\qquad\qquad\qquad u^2 > 19.6$

it follows that: $\qquad\qquad v^2 > 0$

And so the bead will move in complete circles.

(c) The highest point of P's path will be the point where P's speed is zero.

Let the angle between OP and the downward vertical at this point be θ.

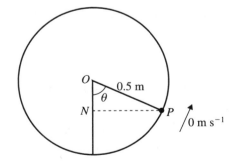

Initial situation $\qquad\qquad\quad$ P at the highest point

Since $ON = 0.5\cos\theta$, the vertical distance P has risen is:

$$0.5 - 0.5\cos\theta = 0.5(1 - \cos\theta)$$

Using the work–energy principle:

$$\tfrac{1}{2}mu^2 - 0 = mgh$$

$$\tfrac{1}{2}m \times 4.9 = m \times 9.8 \times 0.5(1 - \cos\theta)$$

$$4.9 = 9.8 \times (1 - \cos\theta)$$

$$\tfrac{1}{2} = 1 - \cos\theta$$

$$\cos\theta = \tfrac{1}{2}$$

$$\theta = 60°$$

The angle between OP and the downward vertical at the highest point of P's path is 60°.

Exercise 4E

Whenever a numerical value of g is required, take $g = 9.8\,\mathrm{m\,s^{-2}}$.

1 A particle of mass 1.5 kg is attached to one end A of a light rod AB of length 0.75 m which is free to rotate in a vertical plane about the other end B. The rod is hanging with A below B. The particle is projected horizontally with speed $10\,\mathrm{m\,s^{-1}}$. Calculate (a) the speed of the particle when it is vertically above B (b) the tension in the rod at this time.

2 A particle of mass 0.7 kg is attached to one end A of a light rod AB of length 0.5 m which is free to rotate in a vertical plane about the other end B. AB is held in a horizontal position and then released. Calculate the speed of the particle at the lowest point of its path.

3 A pendulum bob has mass 0.2 kg. It is attached to one end of a light rod of length 2 m. The rod is free to rotate in a vertical plane about an axis through the other end O. Given that the pendulum swings through 60° on either side of the vertical, calculate the speed of the bob at the lowest point of its path and the tension in the rod when the rod makes an angle of 30° with the downward vertical.

4 A particle of mass 0.5 kg is attached to one end A of a light rod AB of length 2 m which is free to rotate in a vertical plane

about the other end B. AB hangs vertically with A below B. The particle is projected horizontally with speed $12 \, \text{m s}^{-1}$. (a) Calculate the speed of the particle when AB is horizontal. When AB is horizontal the particle strikes and adheres to an identical particle which is at rest. (b) Calculate the speed of the combined particle and determine whether this combined particle will describe complete circles.

5   A particle P is attached to one end of a light rod of length l which is free to rotate in a vertical plane about its other end. The rod hangs vertically with P below the fixed end. (a) Given that P is projected horizontally with speed $3\sqrt{(lg)}$ prove that P describes complete circles. (b) Find the least value of the speed of projection if P is to describe complete circles.

6   A bead B of mass m is threaded on a smooth circular wire of radius a and centre O which is fixed in a vertical plane. B is released from rest at the point where OB makes an angle of $30°$ with the upward vertical. (a) State how you could model the bead in order to investigate its motion, showing clearly any assumptions made. (b) Given that the speed of B at the lowest point of its path is v, find v^2 in terms of a and g. (c) Find in terms of m and g the magnitude and direction of the reaction of the wire on the bead when OB makes an angle of $60°$ with the upward vertical.

7   A particle P of mass $2 \, \text{kg}$ is attached to one end of a light rod of length $0.5 \, \text{m}$ which is free to rotate in a vertical plane about its other end. The particle describes complete circles. Given that the tension at the lowest point of P's path is five times the tension at the highest point, find the speed of P at its lowest point.

Motion of a particle which can leave the circular path

In each of examples 16 and 17 the particle was unable to leave the circular path on which it was moving. However, if a particle P is attached to one end of a *string* and set in motion in a vertical circle, the motion of the particle will be described by one of three distinct cases.

(1)

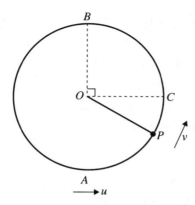

The speed of the particle at the lowest point A is sufficient for the particle to pass through the highest point B of the circular path *with the string still taut*. In this case the particle will describe complete circles.

(2)

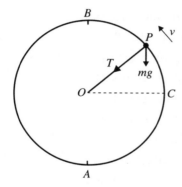

The speed of the particle at A is sufficient for the particle to pass through the point C on the same horizontal level as the centre of the circle but the string becomes slack at some point between C and B. Once the string becomes slack the only force acting on the particle is gravity, but the particle will have a velocity and so will move as a projectile until the string becomes taut once more.

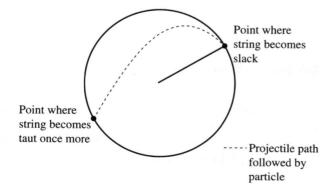

(3) The speed of the particle at A is insufficient for the particle to rise above the level of the centre of the circle. In this case the particle oscillates as a pendulum.

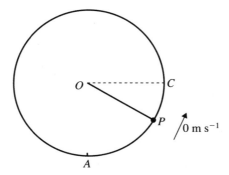

For a particle attached to a string, the motion can only be in a circle when the string is taut.

Hence:

the condition for complete circles is that there is a tension at the highest point of the circular path.

Consider a particle P of mass m attached to one end of a light inextensible string of length r. The other end of the string is fixed at a point O and P hangs freely in equilibrium. Suppose P is projected horizontally with speed u.

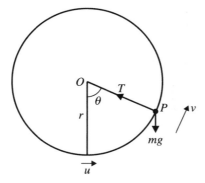

Let the speed of P when OP makes an angle θ with the downward vertical be v and the tension in the string be T.

By the work–energy principle:

$$\text{K.E. lost} = \text{P.E. gained}$$

$$\tfrac{1}{2}mu^2 - \tfrac{1}{2}mv^2 = mgh$$

$$\tfrac{1}{2}mu^2 - \tfrac{1}{2}mv^2 = mgr(1 - \cos\theta)$$

$$u^2 - v^2 = 2gr(1 - \cos\theta)$$

$$v^2 = u^2 - 2gr(1 - \cos\theta) \qquad (1)$$

Using the equation of motion, $F = ma$, along PO, with $a = \dfrac{v^2}{r}$ gives:

$$T - mg\cos\theta = m\frac{v^2}{r} \qquad (2)$$

Eliminating v^2 between equations (1) and (2) gives:

$$T - mg\cos\theta = \frac{m}{r}\left[u^2 - 2gr(1 - \cos\theta)\right]$$

$$T = mg\cos\theta + m\frac{u^2}{r} - 2mg(1 - \cos\theta)$$

$$T = m\left(\frac{u^2}{r} - 2g + 3g\cos\theta\right)$$

Case (1)   For complete circles the string must be taut at the highest point of the path.

That is:

$$T \geqslant 0 \text{ when } \theta = 180°$$

So:

$$m\left(\frac{u^2}{r} - 2g + 3g\cos 180°\right) \geqslant 0$$

$$\frac{u^2}{r} \geqslant 2g - 3g\cos 180°$$

$$\frac{u^2}{r} \geqslant 2g + 3g \qquad \text{as } \cos 180° = -1$$

$$u^2 \geqslant 5rg$$

Case (2)   The string becomes slack when the particle is above the level of O.

That is:

$$T = 0 \text{ for some } \theta \text{ between } 90° \text{ and } 180°$$

From above:

$$T = m\left(\frac{u^2}{r} - 2g + 3g\cos\theta\right)$$

So $T = 0$ when:

$$3g\cos\theta = 2g - \frac{u^2}{r}$$

that is, when:

$$\cos\theta = \frac{2rg - u^2}{3rg}$$

But:

$$90° < \theta < 180° \Rightarrow 0 > \cos\theta > -1$$

So:

$$0 > \frac{2rg - u^2}{3rg} > -1$$

Separating this into two inequalities gives:

$$0 > \frac{2rg - u^2}{3rg} \qquad (3)$$

and:

$$\frac{2rg - u^2}{3rg} > -1 \qquad (4)$$

As $3rg$ must be positive you can multiply these inequalities by $3rg$.

So from (3): $$u^2 > 2rg$$

and from (4): $$2rg - u^2 > -3rg$$

$$5rg > u^2$$

Hence the string becomes slack when the particle is above the level of O if:

$$2rg < u^2 < 5rg$$

Case (3) The particle oscillates like a pendulum if the speed is zero before the particle reaches the level of O.

That is: $$v = 0 \text{ for some } \theta \leqslant 90°$$

From equation (1) above:

$$v^2 = u^2 - 2gr(1 - \cos \theta)$$

So $v = 0$ when: $$u^2 - 2gr + 2gr \cos \theta = 0$$

that is, when: $$\cos \theta = \frac{2gr - u^2}{2gr}$$

But as $0 < \theta \leqslant 90°$ it follows that $\cos \theta \geqslant 0$.

So: $$u^2 \leqslant 2gr$$

Hence, if the speed of the particle at the lowest point of its path is u:

$$u^2 \geqslant 5gr \Rightarrow \text{ the particle will describe complete circles}$$

$$2gr < u^2 < 5gr \Rightarrow \text{ the particle will rise above the level of the} \\ \text{centre of the circle and will leave the circular} \\ \text{path}$$

$$u^2 \leqslant 2gr \Rightarrow \text{ the particle will oscillate as a pendulum.}$$

Example 18

A particle of mass m is attached to one end A of a light inelastic string of length r, the other end B being fixed. Initially the particle hangs vertically below B. It is projected horizontally with speed $2\sqrt{(gr)}$. Show that the particle leaves its circular path when AB makes an angle $\arccos \frac{2}{3}$ with the upward vertical.

The particle will leave its circular path when the string becomes slack.

Let the angle between AB and the upward vertical at this moment be θ. Let the speed of the particle be v and the tension in the string be T.

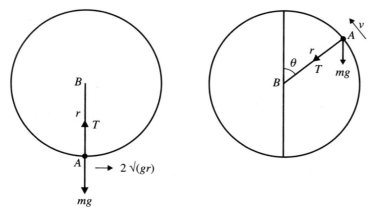

The component of the equation of motion along the inward radius gives:

$$T + mg\cos\theta = m\frac{v^2}{r}$$

When $T = 0$:

$$mg\cos\theta = m\frac{v^2}{r}$$

$$\cos\theta = \frac{v^2}{gr} \qquad (1)$$

Using the work–energy principle gives:

$$\tfrac{1}{2}mu^2 - \tfrac{1}{2}mv^2 = mgr(1 + \cos\theta)$$

As $u = 2\sqrt{(gr)}$:

$$\tfrac{1}{2}m \times 4gr - \tfrac{1}{2}mv^2 = mgr + mgr\cos\theta$$
$$4gr - v^2 = 2gr + 2gr\cos\theta$$
$$v^2 = 2gr - 2gr\cos\theta$$
$$\frac{v^2}{gr} = 2 - 2\cos\theta$$

But from equation (1) above $\dfrac{v^2}{gr} = \cos\theta$

So:

$$\cos\theta = 2 - 2\cos\theta$$
$$3\cos\theta = 2$$
$$\cos\theta = \tfrac{2}{3}$$
$$\theta = \arccos\tfrac{2}{3}$$

The particle will leave its circular path when AB makes an angle $\arccos\tfrac{2}{3}$ with the upward vertical.

A particle which moves on a circular cylinder or sphere, either on the inside or outside of it, will experience a contact force from the cylinder or sphere. When that contact force becomes zero the particle will leave its circular path.

Example 19

A particle P of mass $2\,\mathrm{kg}$ rests at the highest point of a smooth sphere centre O of radius $0.5\,\mathrm{m}$ which is fixed to a horizontal table. P is slightly disturbed and slides on the surface of the sphere.

(a) Determine the angle between OP and the upward vertical when the particle leaves the surface of the sphere.
(b) Calculate the magnitude and direction of the particle's velocity when it hits the table.

(a) Consider the particle when OP has turned through an angle θ. Let the speed of the particle at this moment be $v\,\mathrm{m\,s}^{-1}$. Let the normal contact force between P and the sphere be R N.

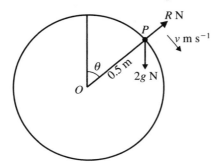

Using the work–energy principle gives:

$$\tfrac{1}{2}mv^2 - \tfrac{1}{2}mu^2 = mgh$$

$$\tfrac{1}{2} \times 2 \times v^2 - 0 = 2 \times 9.8 \times 0.5(1 - \cos\theta)$$

$$v^2 = 9.8(1 - \cos\theta) \tag{1}$$

Using the equation of motion, $F = m\frac{v^2}{r}$, along the inward radius gives:

$$2 \times 9.8 \cos\theta - R = \frac{2v^2}{0.5}$$

$$v^2 = \tfrac{1}{4}(2 \times 9.8 \cos\theta - R) \tag{2}$$

Equating the expressions for v^2 from equations (1) and (2) gives:

$$\tfrac{1}{4}(2 \times 9.8 \cos\theta - R) = 9.8(1 - \cos\theta)$$

The particle leaves the surface when $R = 0$, that is when:

$$\tfrac{1}{2} \times 9.8 \cos\theta = 9.8(1 - \cos\theta)$$

$$\tfrac{1}{2}\cos\theta = 1 - \cos\theta$$

$$\tfrac{3}{2}\cos\theta = 1$$

$$\cos\theta = \tfrac{2}{3}$$

$$\theta = 48.18°$$

The particle leaves the surface when the angle between OP and the upward vertical is $48.2°$.

(b) Once the particle leaves the surface of the sphere the only force acting upon it is gravity. The particle will therefore move as a projectile (see Book M2, section 1.1).

To investigate the motion of this projectile you must first find the velocity of the particle as it leaves the sphere.

From equation (2) above:

$$v^2 = \tfrac{1}{4}(2 \times 9.8 \times \cos\theta - R)$$

When the particle leaves the sphere, $\cos\theta = \tfrac{2}{3}$ and $R = 0$.

So:
$$v^2 = \tfrac{1}{4} \times 2 \times 9.8 \times \tfrac{2}{3}$$

$$v^2 = \frac{9.8}{3}$$

OP makes an angle of $48.18°$ with the upward vertical at the moment the particle leaves the sphere.

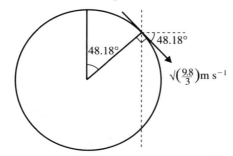

Using geometry, the diagram shows that the particle's velocity is at an angle of $48.18°$ below the horizontal.

So for the projectile motion, the particle's velocity has components:

$$\sqrt{\left(\frac{9.8}{3}\right)}\cos 48.18°\quad\text{horizontally}$$

and:
$$\sqrt{\left(\frac{9.8}{3}\right)}\sin 48.18°\quad\text{vertically downwards}$$

The vertical distance travelled by the particle from leaving the sphere to reaching the table is:

$$r(1 + \cos 48.18°) = 0.5 \times \frac{5}{3} = \frac{2.5}{3}\ \text{m}$$

as $\cos 48.18° = \tfrac{2}{3}$ from part (a).

For the vertical speed, using $v^2 = u^2 + 2as$ with $a = 9.8\ \text{m s}^{-2}$,

$u = \sqrt{\left(\frac{9.8}{3}\right)}\sin 48.18°\ \text{m s}^{-1}$ and $s = \frac{2.5}{3}\ \text{m}$ gives:

$$v^2 = \frac{9.8}{3}\sin^2 48.18° + 2 \times 9.8 \times \frac{2.5}{3}$$

$$v = 4.259\ \text{m s}^{-1}$$

There is no horizontal acceleration and so the horizontal speed is unchanged throughout the projectile motion.

$$\text{Horizontal speed} = \sqrt{\left(\frac{9.8}{3}\right)} \cos 48.18°$$

$$= 1.205 \, \text{m s}^{-1}$$

From the vector diagram, the resultant speed is:

$$\sqrt{\{1.205^2 + 4.259^2\}} = 4.426 \, \text{m s}^{-1}$$

and:

$$\tan \phi = \frac{4.259}{1.205}$$

so:

$$\phi = 74.20°$$

The particle hits the table with a speed of $4.43 \, \text{m s}^{-1}$ at an angle of $74.2°$ below the horizontal.

Example 20

A smooth hemispherical bowl of radius r and centre O is fixed with its rim horizontal. A marble P of mass m is projected vertically downwards inside the bowl from a point on the rim. The initial speed of P is u. When P reaches the lowest point of the bowl it strikes a stationary marble Q of mass $2m$. P is brought to rest by the impact.

(a) Describe the models you would use for P and Q to investigate the subsequent motion, stating clearly any assumptions made.
(b) Show that Q will leave the bowl provided

$$u^2 > 6gr$$

(c) Given that $u^2 = 4gr$, calculate the magnitude of the reaction between Q and the bowl at the highest point of Q's path.

(a) A marble can be assumed to have a small radius compared with the radius of the bowl. P and Q can therefore be modelled as particles.

As the bowl is described as being smooth, it can be assumed that there is no frictional force acting on the marbles.

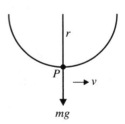

(b) Consider the motion of P from the rim to the lowest point of the bowl. Let the speed of P at the lowest point be v.

$$\text{K.E. gained by } P = \tfrac{1}{2}mv^2 - \tfrac{1}{2}mu^2$$
$$\text{P.E. lost by } P = mgh = mgr$$

By the work–energy principle:

$$\text{K.E. gained} = \text{P.E. lost}$$
$$\tfrac{1}{2}mv^2 - \tfrac{1}{2}mu^2 = mgr$$
$$v^2 = 2gr + u^2$$

Consider the impact of P and Q.

Before impact:

$\longrightarrow \sqrt{(2gr + u^2)}$ $\longrightarrow 0$

P Q

$\bigcirc$ $\bigcirc$

m $2m$

$\longrightarrow 0$ $\longrightarrow V$

After impact:

P Q

$\bigcirc$ $\bigcirc$

m $2m$

Let the speed of Q after impact be V.

By the principle of conservation of momentum (see Book M1, section 5.3):

$$m_1 u_1 + m_2 u_2 = m_1 v_1 + m_2 v_2$$

So:

$$m\sqrt{(2gr + u^2)} + 0 = 0 + 2mV$$

$$V = \tfrac{1}{2}\sqrt{(2gr + u^2)}$$

Consider Q when OQ makes an angle θ with the downward vertical. Let the speed of Q at this point be W.

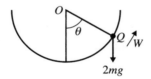

From the lowest point to this level:

$$\text{P.E. gained by } Q = mgh = 2mgr(1 - \cos\theta)$$

$$\text{K.E. lost by } Q = \tfrac{1}{2} \times 2mV^2 - \tfrac{1}{2} \times 2mW^2$$

$$= \tfrac{1}{4}m(2gr + u^2) - mW^2$$

as $V = \tfrac{1}{2}\sqrt{(2gr + u^2)}$.

By the work–energy principle:

$$\text{K.E. lost} = \text{P.E. gained}$$

$$\tfrac{1}{4}m(2gr + u^2) - mW^2 = 2mgr(1 - \cos\theta)$$

$$\tfrac{1}{2}gr + \tfrac{1}{4}u^2 - W^2 = 2gr - 2gr\cos\theta \qquad (1)$$

$$W^2 = \tfrac{1}{4}u^2 - \tfrac{3}{2}gr + 2gr\cos\theta$$

Q will leave the bowl provided $W^2 > 0$ when $\theta = 90°$, that is provided:

$$\tfrac{1}{4}u^2 - \tfrac{3}{2}gr + 0 > 0$$

or:

$$u^2 > 6gr$$

(c) From equation (1) on the previous page:

$$\tfrac{1}{2}gr + \tfrac{1}{4}u^2 - W^2 = 2gr - 2gr\cos\theta$$

As $u^2 = 4gr$ this gives:

$$\tfrac{1}{2}gr + gr - W^2 = 2gr - 2gr\cos\theta$$

At the highest point of Q's path $W = 0$.

So:

$$\tfrac{1}{2}gr + gr = 2gr - 2gr\cos\theta$$

$$2gr\cos\theta = \tfrac{1}{2}gr$$

$$\cos\theta = \tfrac{1}{4}$$

Let the normal reaction between Q and the bowl be R.

Using Newton's second law, $F = ma$, along the radius with $a = \dfrac{W^2}{r}$

gives:

$$R - 2mg\cos\theta = 2m\frac{W^2}{r}$$

but at the highest point of Q's path, $W = 0$ and $\cos\theta = \tfrac{1}{4}$

so:

$$R - 2mg \times \tfrac{1}{4} = 0$$

$$R = \tfrac{1}{2}mg$$

The magnitude of the normal reaction between Q and the bowl is $\tfrac{1}{2}mg$.

Exercise 4F

Whenever a numerical value of g is required, take $g = 9.8\,\mathrm{m\,s}^{-2}$.

1 A light inelastic string of length 1 m has one end fixed to a point O. A particle P of mass 0.25 kg is hanging in equilibrium from the free end of the string. P is projected horizontally with a speed which is just sufficient for P to reach the level of O. Calculate the speed of projection and the tension in the string when OP makes an angle of 60° with the vertical.

2 A light inelastic string of length 2 m has one end fixed to a
   point O. A particle P of mass 1 kg is hanging in equilibrium
   from the free end of the string. P is projected horizontally.
   (a) Calculate the minimum speed of projection if P is to
   describe complete circles.
   (b) Given that P is projected at twice this minimum speed,
   calculate the maximum and minimum magnitudes of the
   tension in the string.

3 A particle P of mass 0.5 kg is hanging in equilibrium from one
   end of a light inelastic string of length 1.5 m whose other end
   is fixed at a point O. P is projected horizontally with speed
   u m s^{-1}. (a) Given that the string becomes slack when OP
   makes an angle of 45° with the upward vertical, calculate the
   value of u. (b) Describe briefly the subsequent motion of the
   particle.

4 A particle P of mass 0.5 kg is hanging in equilibrium from one
   end of a light inelastic string of length 1.5 m whose other end
   is fixed at a point O. P is projected horizontally with speed
   u m s^{-1}. When OP is horizontal it meets a small smooth peg at
   Q where $OQ = 1$ m.
   (a) Calculate the minimum value of u if P is to describe a
   complete circle about Q.
   (b) Given that P is projected with $\frac{9}{10}$ of this speed, calculate the
   angle between PQ and the vertical when the string becomes
   slack.

5 A child of mass 20 kg is sitting on a swing which has ropes of
   length 2.5 m. She is pulled back until the ropes make an angle
   of 50° with the downward vertical and then released.
   (a) State how you would model the child in order to
   investigate the subsequent motion.
   (b) Calculate the speed with which the child passes through
   the lowest point of the swing and calculate the tension in each
   of the two ropes at this point.

6 A light inelastic string of length l has one end fixed to a fixed
   point O. A particle P of mass m is attached to the other end.
   P is held vertically above O with the string taut and is
   projected horizontally with speed $\sqrt{(gl)}$.

(a) Find, in terms of l, m, g and θ the tension in the string when OP makes an angle θ with the upward vertical.

(b) Given that the string will break when the tension is $4mg$, find the angle between OP and the upward vertical when the string breaks.

7 A particle of mass m is attached to end A of a light inelastic string AB of length l which is fixed at B. The particle is held with AB horizontal and is projected vertically with speed u. Show that the least value of u for the particle to describe complete circles is independent of the direction of projection. Calculate this least value.

8 A particle P of mass m is at rest at the lowest point of a smooth hemispherical bowl of radius r and centre O whose rim is horizontal. P is projected horizontally with speed u.

(a) If P oscillates through $60°$ on either side of the vertical, find u in terms of r and g and find the magnitude of the reaction between P and the bowl at the highest point of P's path.

(b) If P is about to lose contact with the bowl when OP is horizontal, find u in terms of r and g.

9 A smooth hemispherical bowl of radius a has its rim horizontal. A particle P is projected downwards with speed u from a point on the surface of the bowl $\frac{1}{2}a$ below the level of the rim. If P remains in contact with the bowl during the subsequent motion show that $u \leqslant \sqrt{(ag)}$.

If $u = \sqrt{(2ag)}$, find the speed of P as it leaves the bowl.

10 A particle P rests at the highest point of a fixed smooth sphere of radius a and centre O. P is projected horizontally with speed u. Given that P leaves the surface of the sphere when OP makes an angle θ with the upward vertical, show that:

$$\cos \theta = \frac{2ga + u^2}{3ga}$$

Hence show that if $u \geqslant \sqrt{(ag)}$ the particle will leave the surface of the sphere the instant it is projected.

11 A smooth hemispherical bowl with centre O and radius a is fixed on a horizontal surface so that its rim is in a vertical plane. A small bead is projected horizontally with speed u

from the lowest point A of the rim so that it moves in a vertical circle with centre O. When the bead reaches the highest point B of the rim it is on the point of losing contact with the bowl. (a) State the mathematical model you could use for the bead in order to investigate its motion. (b) Find the speed of the bead at B in terms of a and g. (c) Find u in terms of a and g.

The bead strikes the horizontal surface again at C. (d) Find the distance AC. (e) State an assumption you have made about the forces acting during this motion.

12 A bowl is made from a smooth spherical shell of radius r by cutting away the part which is more than $\frac{1}{3}r$ above the horizontal plane through its centre. A marble is projected from the lowest point of the bowl with speed u. Show that in the subsequent motion the marble will leave the bowl and not fall back into it provided:

$$u^2 > \frac{17gr}{3}$$

Exercise 4G   Mixed questions

Whenever a numerical value of g is required, take $g = 9.8\,\mathrm{m\,s^{-2}}$.

1 An small ball B of mass $2\,\mathrm{kg}$ is attached to one end of a light inextensible string of length $0.25\,\mathrm{m}$. The other end of the string is fixed to a point A of a smooth horizontal table. B moves on the table in a circle, centre A, with constant speed. Given that the string will break when the tension exceeds $10\,\mathrm{N}$, find the maximum angular speed of B.

2 A particle P of mass m is at rest at the highest point of a fixed smooth sphere of radius r and centre A. P receives an impulse of magnitude $2mu$.
(a) Determine the speed with which P begins to move.
P slides on the outer surface of the sphere and loses contact with the sphere when AP makes an angle α with the upward vertical.
(b) Find, in terms of u, r and g, an expression for $\cos\alpha$.
(c) Deduce that $u^2 \leqslant \frac{1}{4}rg$.

3 A car is travelling at a constant speed of $16\,\mathrm{m\,s^{-1}}$ round a bend in a road which is banked at an angle θ to the horizontal. The car is modelled as a particle moving in a horizontal circle of radius 80 m. Assuming there is no sideways frictional force on the car, calculate, in degrees to one decimal place, the value of θ.

4 A hollow cylinder of internal radius 0.5 m is fixed with its axis horizontal. The point A is on the axis of the cylinder and the point B is vertically below A on the internal surface of the cylinder. A particle P is initially at rest at the point B. The particle receives an impulse and moves in a vertical circle, centre A, with initial speed $4\,\mathrm{m\,s^{-1}}$. The particle leaves the surface of the sphere at the point C. Find
(a) the angle, in degrees, between AC and the upward vertical
(b) the speed of P at C
(c) the greatest height above A reached by P.

5 A small brick B of mass 1.5 kg is attached to one end of light rope of length 0.75 m. The other end of the rope is attached to a fixed point A. The brick is moving with constant speed in a horizontal circle with centre vertically below A. The rope is inclined at an angle α to the downward vertical where $\cos\alpha = \frac{4}{5}$. In an initial model of the situation, the brick is modelled as a particle and the rope is assumed to be inextensible. Calculate
(a) the tension in the rope
(b) the angular speed of the brick.
In a refinement of the model, the brick is again modelled as a particle and the rope is inclined at the same angle to the vertical but the rope is modelled as an elastic string of natural length 0.75 m and modulus of elasticity 1470 N. Using this model,
(c) calculate the stretched length of the rope.

6 A boy is cycling round a rough circular path of radius 3 m with constant speed. The path is banked at an angle θ to the horizontal, where $\tan\theta = \frac{7}{24}$ and the coefficient of friction between the path and the cycle is $\frac{3}{7}$. By modelling the boy and his cycle as a particle and assuming that the resistance from non-gravitational forces is negligible, calculate, in $m\,s^{-1}$, the maximum speed at which the boy can cycle without slipping sideways.

7 A conker of mass m is attached to one end of a rod of length l. The other end of the rod is attached to a fixed point O. The conker moves in a vertical circle with centre O. The rod can be assumed to be light and inextensible and the conker can be modelled as a particle. The minimum speed of the conker is a quarter of its maximum speed. All forces acting on the conker, apart from its weight and the force in the rod, may be assumed to be negligible.

(a) Find, in terms of g and l, the speed of the conker when the rod is at an angle of 60° to the upward vertical.

(b) Find, in terms of m and g, the magnitude of the force in the rod when it is at an angle of 60° to the upward vertical.

(c) State whether the force in the rod is a tension or a thrust.

(d) State one physical factor that has been ignored in your model.

8 It is required to estimate the force exerted by a hammer thrower. The athlete is regarded as the vertical axis of a conical pendulum. The hammer is treated as a particle of mass 6 kg attached to a light inextensible wire of length 1 m. The length of the wire together with the athlete's arm is estimated to be 2 m. The particle has angular speed $5\,rad\,s^{-1}$.

(a) Calculate the magnitude of the force exerted by the hammer thrower.

(b) Find the angle between the wire and the vertical.

(c) Discuss briefly the assumptions made in this model.

SUMMARY OF KEY POINTS

1 When a particle P is moving on a circle centre O and the angle between OP and a fixed radius OA is θ, the angular speed of OP is ω where

$$\omega = \frac{d\theta}{dt}$$

2 Angular speed is usually measured in radians per second (rad s^{-1}) or revolutions per minute (rev min^{-1}).

3 The linear speed v of the particle is directed along the tangent to the circular path.

4 The linear and angular speeds are connected by the relationship

$$v = r\omega$$

where r is the radius of the circular path.

5 When r is measured in metres and ω in radians per second, v is in metres per second.

6 A particle moving in a circular path at a constant speed has an acceleration of magnitude a directed towards the centre of the circle where a is given by

$$a = r\omega^2 \text{ or } a = \frac{v^2}{r}$$

7 The resultant force $\mathbf{F}$ acting on a particle of mass m moving on a circular path with constant speed has magnitude $mr\omega^2$ or $\dfrac{mv^2}{r}$ and is directed towards the centre of the circle.

8 A surface which is inclined at an angle to the horizontal is said to be *banked*.

Motion of a particle in a vertical circle

9 A particle which is moving in a *vertical* circle has variable speed.

10 The acceleration of a particle moving in a circle of radius r with speed v at time t has two components:

$$\frac{v^2}{r} \text{ towards the centre of the circle}$$

and: $\quad \dfrac{dv}{dt}$ along the tangent.

11 The component of the force along the radius does no work during the motion.

12 By the work–energy principle, the sum of the K.E. and P.E. of the particle is constant throughout the motion.

13 A particle which cannot leave its vertical circular path (for example a particle on a rod) will describe complete circles provided its velocity at the highest point of the circle is greater than or equal to zero.

14 A particle which can leave its vertical circular path (for example a particle on a string) will do so when the force towards the centre of the circle becomes zero.

Statics of rigid bodies

5

5.1 Centre of mass of a uniform plane lamina

In chapter 2 of Book M2 the centre of mass of a uniform plane lamina was determined using its symmetries. Now a more general method of determining the position of the centre of mass is required which can be used when the lamina has no symmetries. This more general method uses the ideas of calculus.

The following 'key points' from chapter 2 of Book M2, will be required:

1 The centre of mass of a lamina is the point at which the weight acts.

2 The weight of a uniform lamina is evenly distributed throughout its area.

3 The centre of mass of a lamina must lie on any axis of symmetry.

4 The centre of mass of a system of n masses $m_1, m_2, ..., m_n$ at points with coordinates (x_1, y_1), (x_2, y_2), ..., (x_n, y_n) has coordinates $(\bar{x}, \bar{y})$

where: $\qquad \bar{x} = \dfrac{\sum m_i x_i}{\sum m_i} \quad$ and $\quad \bar{y} = \dfrac{\sum m_i y_i}{\sum m_i}$

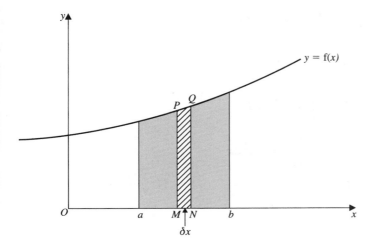

Consider a uniform lamina bounded by the curve with equation $y = f(x)$, the x-axis and the lines $x = a$ and $x = b$ (the region lightly shaded in the figure). Let the mass per unit area of the lamina be ρ (the Greek lower case letter 'rho'). Consider the lamina as made up of elementary strips such as $MNQP$ each parallel to the y-axis. Let $OM = x$ and $MN = \delta x$.

Then:

length of elementary strip $MNQP = y$

area of elementary strip $= y\delta x$

mass of elementary strip $= \rho y \delta x$

coordinates of centre of mass of strip $\simeq (x, \tfrac{1}{2}y)$

The x-coordinate of the centre of mass of the lamina, $\bar{x}$, is now obtained from:

$$\bar{x} = \frac{\sum\limits_i m_i x_i}{\sum\limits_i m_i} \simeq \frac{\sum\limits_{x=a}^{x=b} (\rho y \delta x)x}{\sum\limits_{x=a}^{x=b} (\rho y \delta x)}$$

Taking the limit as $\delta x \to 0$ the summations become integrals. So for the lamina shown in the figure:

■
$$\bar{x} = \frac{\int_a^b x(\rho y \, dx)}{\int_a^b \rho y \, dx} = \frac{\int_a^b x f(x) \, dx}{\int_a^b f(x) \, dx}$$

To find the y-coordinate of the centre of mass of the lamina it is necessary to note that the centre of mass of the strip has y-coordinate equal to $\tfrac{1}{2}y$. Then:

■
$$\bar{y} = \frac{\int_a^b \tfrac{1}{2}y(\rho y \, dx)}{\int_a^b \rho y \, dx} = \frac{\tfrac{1}{2}\int_a^b y^2 \, dx}{\int_a^b y \, dx} \qquad \text{where } y = f(x).$$

These formulae are often written in the alternative form:

$$\left[\int_a^b \rho y \, dx\right]\bar{x} = M\bar{x} = \int_a^b \rho x y \, dx = \int_a^b \rho x \, f(x) \, dx$$

$$\left[\int_a^b \rho y \, dx\right]\bar{y} = M\bar{y} = \tfrac{1}{2}\int_a^b \rho y^2 \, dx = \tfrac{1}{2}\int_a^b \rho[f(x)]^2 \, dx$$

where M is the total mass of the lamina.

Example 1

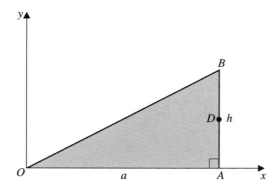

Find the coordinates of the centre of mass of the triangular lamina OAB shown in the figure where $OA = a$ and $AB = h$.

The equation of OB is:
$$y = \frac{h}{a}x$$

The mass of the lamina is:
$$\rho \text{ (area)} = \rho\left(\tfrac{1}{2}ah\right) = M$$

Using the formula for $\bar{x}$:

$$M\bar{x} = \int_0^a \rho xy\,dx = \int_0^a \rho x\frac{h}{a}x\,dx = \frac{h}{a}\rho\int_0^a x^2\,dx$$

$$= \frac{h}{a}\rho\left[\frac{x^3}{3}\right]_0^a = \frac{1}{3}ha^2\rho$$

So:
$$\bar{x} = \frac{\frac{1}{3}ha^2\rho}{\frac{1}{2}ha\rho} = \tfrac{2}{3}a$$

Similarly :
$$M\bar{y} = \int_0^a \rho\tfrac{1}{2}y^2\,dx = \tfrac{1}{2}\rho\int \frac{h^2x^2}{a^2}\,dx$$

$$= \frac{\rho h^2}{2a^2}\left[\frac{x^3}{3}\right]_0^a = \frac{\rho h^2 a}{6}$$

So:
$$\bar{y} = \frac{\frac{1}{6}\rho h^2 a}{\frac{1}{2}\rho ha} = \tfrac{1}{3}h$$

The centre of mass G of the lamina has coordinates $\left(\tfrac{2}{3}a, \tfrac{1}{3}h\right)$.

This shows that if D is the mid-point of AB then $GD = \tfrac{1}{3}OD$ as stated in section 2.2 of Book M2.

Example 2

Find the coordinates of the centre of mass of the uniform lamina bounded by the curve with equation $y^2 = 4x$, the x-axis and the line $x = 1$.

First sketch the lamina:

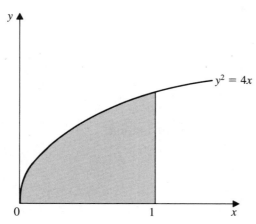

The mass of the lamina is:

$$M = \int_0^1 \rho y \, dx = \rho \int_0^1 2x^{\frac{1}{2}} \, dx$$

$$= 2\rho \left[\tfrac{2}{3} x^{\frac{3}{2}} \right]_0^1 = 2\rho \times \tfrac{2}{3} = \tfrac{4}{3}\rho$$

where ρ is the mass per unit area of the lamina.

The x-coordinate of the centre of mass $\bar{x}$ is obtained from:

$$M\bar{x} = \int_0^1 \rho xy \, dx = \rho \int_0^1 x 2x^{\frac{1}{2}} \, dx = 2\rho \int_0^1 x^{\frac{3}{2}} \, dx$$

$$= 2\rho \left[\tfrac{2}{5} x^{\frac{5}{2}} \right]_0^1 = \tfrac{4}{5}\rho$$

So:

$$\bar{x} = \frac{\frac{4}{5}\rho}{\frac{4}{3}\rho} = \tfrac{3}{5}$$

To find $\bar{y}$ use:

$$M\bar{y} = \int_0^1 \rho \tfrac{1}{2} y^2 \, dx = \tfrac{1}{2}\rho \int_0^1 4x \, dx$$

$$= 2\rho \left[\frac{x^2}{2} \right]_0^1 = 2\rho \tfrac{1}{2} = \rho$$

So:

$$\bar{y} = \frac{\rho}{\frac{4}{3}\rho} = \tfrac{3}{4}$$

The coordinates of the centre of mass are $(\tfrac{3}{5}, \tfrac{3}{4})$.

Example 3

Show that the centre of mass of a uniform semicircular lamina lies on its axis of symmetry at a distance of $\dfrac{4r}{3\pi}$ from the bounding diameter.

Take the bounding diameter as the y-axis and the centre O as the origin. If the point A, on the curved boundary, has coordinates (x, y) then by Pythagoras' theorem $x^2 + y^2 = r^2$.

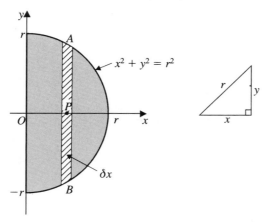

Consider the elementary strip AB of width δx, parallel to the y-axis and at a distance of x from it. The centre of mass of this strip is at the point P with coordinates $(x, 0)$.

The centres of mass of all such strips also lie on the x-axis.

Hence the centre of mass of the lamina will also lie on the x-axis – the axis of symmetry of the lamina.

By Pythagoras' theorem:

$$AP = \sqrt{(r^2 - x^2)}$$

The area of the strip AB is therefore $2\sqrt{(r^2 - x^2)}\delta x$ and its mass is $2\rho\sqrt{(r^2 - x^2)}\delta x$ where ρ is the mass per unit area of the lamina.

The mass of the lamina is $M = \frac{1}{2}\pi r^2 \rho$ and $\bar{x}$ is obtained from:

$$M\bar{x} = \int_0^r x 2\rho\sqrt{(r^2 - x^2)}\,\mathrm{d}x$$

$$= 2\rho \int_0^r x(r^2 - x^2)^{\frac{1}{2}}\,\mathrm{d}x$$

$$= 2\rho \left[-\tfrac{2}{6}\left(r^2 - x^2\right)^{\frac{3}{2}} \right]_0^r$$

$$= \tfrac{2}{3}\rho r^3$$

So:

$$\bar{x} = \frac{\tfrac{2}{3}\rho r^3}{\tfrac{1}{2}\rho\pi r^2} = \frac{4r}{3\pi}$$

That is, the centre of mass is at a distance $\dfrac{4r}{3\pi}$ from the bounding diameter.

Example 4

Show that the centre of mass of a uniform wire in the form of an arc of a circle, radius r, subtending an angle 2α at the centre is at a distance $\dfrac{r\sin\alpha}{\alpha}$ from the centre.

The bisector of the angle 2α is an axis of symmetry of the wire. **The centre of mass will therefore lie on this line.** Take this line to be the x-axis and the centre O of the circle to be the origin.

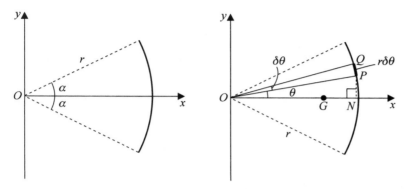

Let ω be the mass per unit length of the wire.

A typical element PQ of the wire subtends an angle $\delta\theta$ at O and has length $r\delta\theta$.

The mass of this element is $\omega r\delta\theta$ and from the diagram the length ON is $r\cos\theta$. (Here N is the foot of the perpendicular from P on to Ox.)

The length of the wire is $r2\alpha$ (see section 2.2 of Book P1) so the mass M of the wire is

$$M = (r2\alpha)\omega$$

The x-coordinate, $\bar{x}$, of the centre of mass is obtained from:

$$M\bar{x} = \sum_{\theta=-\alpha}^{\theta=\alpha}(\omega r\delta\theta)r\cos\theta$$

Taking the limit as $\delta\theta \to 0$ the summation becomes an integral and so:

$$M\bar{x} = \omega r^2 \int_{-\alpha}^{\alpha}\cos\theta\,\delta\theta = \omega r^2[\sin\theta]_{-\alpha}^{\alpha}$$

$$= 2\omega r^2\sin\alpha$$

and:
$$\bar{x} = \frac{2\omega r^2\sin\alpha}{2\omega r\alpha} = \frac{r\sin\alpha}{\alpha}$$

The centre of mass of the wire is at a distance $\dfrac{r\sin\alpha}{\alpha}$ from the centre and lies on the axis of symmetry.

> Note: This result is given in the formula sheet. You will not be expected to reproduce this proof in your M3 exam.

Example 5

Find the centre of mass of a uniform lamina in the form of a sector of a circle, radius r, subtending an angle 2α at the centre.

As in the previous example the bisector of the angle 2α is an axis of symmetry. Choose axes as before with the centre O as the origin and the line of symmetry as the x-axis.

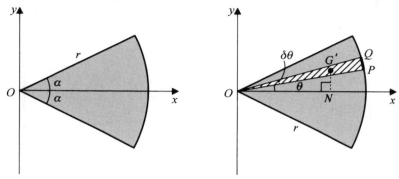

Let ρ be the mass per unit area.

Divide the lamina into elements of the form OPQ, which is a sector subtending an angle of $\delta\theta$ at O (see diagram).

The area of OPQ is $\frac{1}{2}r^2\delta\theta$. (See section 2.3 of Book P1.) Therefore the mass of OPQ is $\frac{1}{2}\rho r^2\delta\theta$.

The sector OPQ is approximately a triangle, so its centre of mass G' is at a distance $\frac{2}{3}r$ from O so that $ON = \frac{2}{3}r\cos\theta$ (see diagram).

The mass M of the whole sector is given by:

$$M = \rho \times \text{(area of whole sector)}$$
$$= \frac{1}{2}\rho r^2 2\alpha = \rho r^2\alpha$$

The x-coordinate, $\bar{x}$, of the centre of mass is obtained from:

$$M\bar{x} = \sum_{\theta=-\alpha}^{\theta=\alpha}\left(\tfrac{1}{2}\rho r^2\delta\theta\right)\tfrac{2}{3}r\cos\theta$$

Taking the limit as $\delta\theta \to 0$ the summation becomes an integral and so:

$$M\bar{x}\int_{-\alpha}^{\alpha}\tfrac{1}{3}\rho r^3\cos\theta\,\mathrm{d}\theta = \tfrac{1}{3}\rho r^3[\sin\theta]_{-\alpha}^{\alpha}$$
$$= \tfrac{1}{3}\rho r^3 2\sin\alpha = \tfrac{2}{3}\rho r^3\sin\alpha$$

and:
$$\bar{x} = \frac{\frac{2}{3}\rho r^3\sin\alpha}{\rho r^2\alpha} = \frac{2r\sin\alpha}{3\alpha}$$

The centre of mass is at a distance $\dfrac{2r\sin\alpha}{3\alpha}$ from the centre of the sector and lies on the axis of symmetry.

In the following example a lamina with a more complicated shape is considered. Such a lamina may be considered as a combination of several simple laminae such that:

(i) the mass of each is easily obtained,

(ii) the position of the centre of mass of each is given in the list of standard results at the end of this chapter and also in the formula book.

If integration is not explicitly asked for in the question you may, in general, quote these standard results.

Example 6

The figure shows a badge cut from a uniform sheet of metal. The badge is formed from two semicircles, each of radius 1 cm and a semicircle of radius 2 cm joined as shown in the figure to make a plane lamina. The point O is the point of contact of the smaller semicircles. Determine, in terms of π, the distance from O of the centre of mass.

The line through O perpendicular to the joining line OX is a line of symmetry and so the centre of mass G lies on this line. Let the mass per unit area be ρ. Then:

mass of smaller semicircle is $\quad \frac{1}{2}\pi(1)^2\rho = \frac{1}{2}\pi\rho$

mass of larger semicircle is $\quad \frac{1}{2}\pi(2)^2\rho = 2\pi\rho$

So: mass of badge is $\quad 2\pi\rho + 2(\frac{1}{2}\pi\rho) = 3\pi\rho$

Using the standard result for the centre of mass of a semicircular lamina, $\frac{4r}{3\pi}$ (see the table at the end of this chapter or the formula book), the centre of mass of a smaller semicircle is a distance $\frac{4}{3\pi}$ **above** OX.

The centre of mass of the larger semicircle is a distance $\frac{4 \times 2}{3\pi} = \frac{8}{3\pi}$ **below** OX.

Let the centre of mass of the badge be a distance $\bar{x}$ **below** OX. Then, as in chapter 2 of Book M2,

$$3\pi\rho\bar{x} = 2\pi\rho\left(\frac{8}{3\pi}\right) - \frac{1}{2}\pi\rho\left(\frac{4}{3\pi}\right) - \frac{1}{2}\pi\rho\left(\frac{4}{3\pi}\right)$$

So: $$\bar{x} = \frac{4}{3\pi}$$

The distance from O of the centre of mass is $\frac{4}{3\pi}$.

Exercise 5A

1 Find the centre of mass of a uniform triangular lamina bounded by the lines $3y = 4x$, $y = 0$ and $x = 3$.

2 Use integration to find the centre of mass of a uniform lamina bounded by the curve with equation $y = x^2$, the x-axis and the line $x = 1$.

3 Use integration to find the centre of mass of a uniform lamina bounded by the curve with equation $y = x^3$, the x-axis and the lines $x = 1$ and $x = 2$.

4 Use integration to find the centre of mass of a uniform lamina bounded by the curve with equation $y^2 = 4ax$ and the line $x = b\,(b > 0)$.

5 Use integration to find the centre of mass of a uniform triangular lamina with vertices $(0, h)$, $(0, 0)$ and $(a, 0)$.

6 Find the centre of mass of a uniform lamina bounded by the curve with equation $y = x^2 + 1$, the x-axis and the lines $x = 0$ and $x = 1$.

7
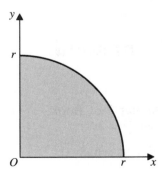

Find the centre of mass of a uniform lamina in the shape of the quadrant of a circle of radius r shown in the figure.

8 Find the centre of mass of a uniform lamina bounded by the curve with equation $y = \cos x \, (x > 0)$ and the lines $x = 0$ and $y = 0$.

9 Find the centre of mass of a uniform lamina bounded by the curve with equation $y = e^x$, the x-axis and the lines $x = 0$ and $x = 1$.

In questions 10 and 11 use the 'subtraction method', that is consider the given lamina as the difference between two laminae.

10

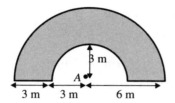

The figure shows a uniform semicircular lamina of radius 6 m from which a semicircular portion of radius 3 m has been removed. Find the distance of the centre of mass of the remaining shape from the point A shown in the figure.

11

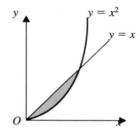

The figure shows a uniform lamina bounded by the curve with equation $y = x^2$ and the line with equation $y = x$. Find the coordinates of the centre of mass of the lamina.

5.2 Centre of mass of a uniform solid body

For a uniform solid body the 'key points' from chapter 2 of Book M2 for a uniform lamina may be generalised so that:

1 The centre of mass of a solid body is the point at which the weight acts.

2 The weight of a uniform solid body is evenly distributed throughout its volume.

3 ● The centre of mass of a uniform solid body must lie on any
axis of symmetry.
  ● The centre of mass of a uniform solid body must lie on any
plane of symmetry.

4 The centre of mass of a system of n masses $m_1, m_2, \ldots, m_n$ at
points with coordinates $(x_1, y_1, z_1), (x_2, y_2, z_2), \ldots, (x_n, y_n, z_n)$ has
coordinates $(\bar{x}, \bar{y}, \bar{z})$ where:

$$\bar{x} = \frac{\sum m_i x_i}{\sum m_i}$$

$$\bar{y} = \frac{\sum m_i y_i}{\sum m_i}$$

$$\bar{z} = \frac{\sum m_i z_i}{\sum m_i}$$

Use of symmetry

The position of the centre of mass of certain uniform solids may be
obtained using just the symmetry properties.

Uniform solid sphere
A uniform solid sphere has an infinite number of planes of
symmetry – every plane through the centre of the sphere is a plane
of symmetry. Since each of these contains the centre G this is
therefore the centre of mass.

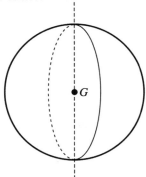

Uniform solid right circular cylinder
The cross-section of a solid right circular cylinder is a circle. The
cylinder therefore has an infinite number of planes of symmetry,
each of which contains the axis of the cylinder. Hence the centre of
mass of a uniform right circular cylinder must lie on the axis.

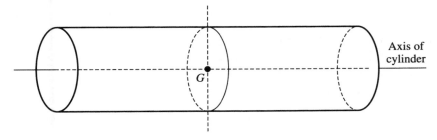

The cylinder also has a plane of symmetry which bisects the axis of the cylinder and is parallel to the plane ends of the cylinder. The centre of mass must also lie on this plane. This plane intersects the axis of symmetry at the point G mid-way between the ends of the cylinder. The point G is the centre of mass of the cylinder.

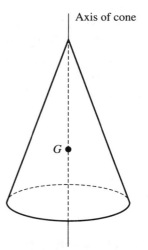

Axis of cone

Uniform solid right circular cone
The axis of a solid right circular cone is an axis of symmetry of the cone and every cross-section perpendicular to this axis is a circle with centre on this line. Hence the centre of mass G of a uniform circular cone lies on the axis of the cone. The exact position of G on the axis will be determined later in this chapter.

Uniform solid hemisphere
Suppose O is the centre of the plane face of the hemisphere. The line through O perpendicular to the plane face is an axis of symmetry. Hence the centre of mass of a uniform solid hemisphere lies on this line. The exact position of G on this line will be found later in this chapter.

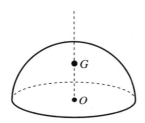

Centres of mass of solids of revolution

Finding the centre of mass of an irregular solid is very difficult. An experimental method may be the only possible approach. There is however a special type of solid known as a **solid of revolution** for which there is a general method using calculus.

A solid of revolution is formed by rotating a region bounded by a known curve through $360°$ about a fixed axis. Examples of such solids are:

   hemisphere, obtained by rotating a quarter circle

   right circular cone, obtained by rotating a triangle

   paraboloid, obtained by rotating a parabola.

As in section 5.1, consider the region bounded by the curve with equation $y = f(x)$, the x-axis and the lines $x = a$ and $x = b$. This is rotated through $360°$ about the x-axis to produce a solid of revolution bounded by $x = a$ and $x = b$. A cross-section of the solid produced is shown in the figure.

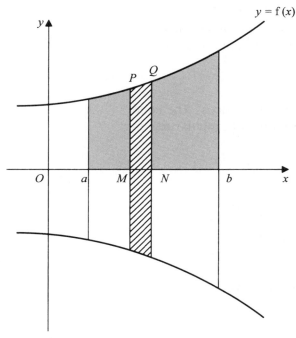

Consider a typical element of the solid formed by rotating the strip $MNQP$ about the x-axis where $OM = x$ and $MN = \delta x$. This typical element is a very thin cylinder whose cross-section is a circle of radius $MP = y$ and area πy^2

So: volume of element $= \pi y^2 \delta x$

mass of element $= \rho \pi y^2 \delta x$

where ρ is the mass per unit volume.

By symmetry the centre of mass of this element lies on the x-axis. Its x-coordinate is equal to $OM = x$.

The x-coordinate of the centre of mass of the whole solid, $\bar{x}$, is now obtained from:

$$\bar{x} = \frac{\sum m_i x_i}{\sum m_i} = \frac{\sum\limits_{x=a}^{x=b} (\rho \pi y^2 \delta x) x}{\sum\limits_{x=a}^{x=b} (\rho \pi y^2 \delta x)}$$

Taking the limit as $\delta x \to 0$, the summations become integrals. So for the solid shown in the figure:

$$\bar{x} = \frac{\int_a^b x \rho \pi y^2 \, \mathrm{d}x}{\int_a^b \rho \pi y^2 \, \mathrm{d}x}$$

or:

$$M\bar{x} = \left[\int_a^b \rho \pi y^2 \, \mathbf{d}x \right] \bar{x} = \int_a^b \rho \pi y^2 x \, \mathbf{d}x$$

Example 7

A uniform solid right circular cone has base radius a and height h. Show that the centre of mass lies on the axis of the cone at a distance $\dfrac{h}{4}$ from the plane base.

Assume the axis of symmetry of the cone is the x-axis. The solid right circular cone may be generated by rotating a triangular region about the x-axis as shown in the figure.

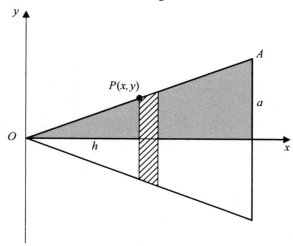

To find the equation of the line OA let the point $P(x, y)$ be on this line. The gradient of the line is $\dfrac{a}{h}$ and so:

$$\frac{y}{x} = \frac{a}{h}$$

or:

$$y = \frac{ax}{h}$$

The shaded triangular region is therefore bounded by the line $y = \dfrac{ax}{h}$, the x-axis and the line $x = h$.

The mass M of the cone is given by:

$$M = \text{volume of cone} \times \rho = \tfrac{1}{3}\pi a^2 h \rho$$

Then:

$$M\bar{x} = \int_0^h x \rho \pi y^2 \, \mathrm{d}x \quad \text{with } y = \frac{a}{h} x$$

$$= \pi \rho \int_0^h x \frac{a^2 x^2}{h^2} \, \mathrm{d}x = \pi \frac{\rho a^2}{h^2} \int_0^h x^3 \, \mathrm{d}x$$

$$= \pi \frac{\rho a^2}{h^2} \left[\frac{x^4}{4} \right]_0^h = \pi \frac{\rho a^2}{h^2} \frac{h^4}{4} = \tfrac{1}{4}\pi \rho a^2 h^2$$

So:
$$\bar{x} = \frac{\frac{1}{4}\pi\rho a^2 h^2}{\frac{1}{3}\pi\rho a^2 h} = \frac{3}{4}h$$

and the centre of mass is a distance of $\frac{3}{4}h$ from O or $\frac{1}{4}h$ from the plane base. As discussed earlier, the centre of mass lies on the axis of the cone as this is an axis of symmetry.

Example 8

Show that the centre of mass of a solid hemisphere of radius a is at a distance $\dfrac{3a}{8}$ from its plane face.

The hemisphere may be obtained by rotating a quarter circle of radius a and centre the origin O, in the 1st quadrant, about the x-axis.

If the coordinates of any point P on the curved boundary are (x, y) then by Pythagoras' theorem $x^2 + y^2 = a^2$. The shaded region is bounded by this curve, the x-axis and the y-axis.
The mass M of the hemisphere is given by:

$$M = \text{volume of hemisphere} \times \rho = \left(\tfrac{2}{3}\pi a^3\right)\rho$$

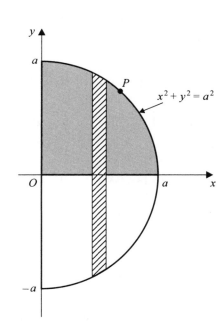

Then:
$$M\bar{x} = \int_0^a x\rho\pi y^2 \, dx \qquad \text{with } y^2 = a^2 - x^2$$

$$= \pi\rho \int_0^a x(a^2 - x^2)\, dx = \pi\rho \int_0^a (a^2 x - x^3)\, dx$$

$$= \pi\rho \left[a^2\frac{x^2}{2} - \frac{x^4}{4} \right]_0^a$$

$$= \pi\rho \left[\frac{a^4}{2} - \frac{a^4}{4} \right] = \frac{\pi}{4}\rho a^4$$

So:
$$\bar{x} = \frac{\frac{\pi}{4}\rho a^4}{\frac{2\pi}{3}\rho a^3} = \frac{3a}{8}$$

The centre of mass is at a distance $\dfrac{3a}{8}$ from the plane face.

Example 9

The region R is bounded by the curve with equation $y^2 = x$, the x-axis and the lines $x = 2$, $x = 4$, and lies in the 1st quadrant. The region is rotated about the x-axis through one revolution. Find the distance of the centre of mass of the solid of revolution from O.

The region R is shown shaded in the figure together with the solid of revolution formed.

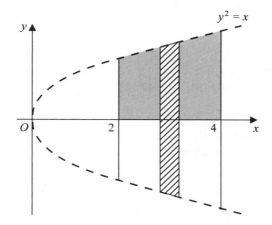

The x-axis is an axis of symmetry of the solid and so the centre of mass lies on this axis.

The mass of the solid of revolution M is given by:

$$M = \rho \times \text{(volume of solid of revolution)}$$

$$M = \rho \int_2^4 \pi y^2 \, dx \qquad \text{where } y^2 = x \text{ (See chapter 9 of Book P2)}$$

$$= \rho \pi \int_2^4 x \, dx$$

$$= \rho \pi \left[\frac{x^2}{2} \right]_2^4$$

$$= \rho \pi \left[\frac{(4)^2}{2} - \frac{(2)^2}{2} \right]$$

$$= \rho \pi [8 - 2]$$

$$= 6 \rho \pi$$

Then: $\quad M\bar{x} = \displaystyle\int_{2}^{4} x\rho\pi y^2 \,\mathrm{d}x \qquad$ where $y^2 = x$

$$= \pi\rho \int_{2}^{4} xx\,\mathrm{d}x = \pi\rho \int_{2}^{4} x^2\,\mathrm{d}x$$

$$= \pi\rho \left[\frac{x^3}{3}\right]_{2}^{4} = \frac{\pi\rho}{3}\left[(4)^3 - (2)^3\right] = \frac{\pi\rho}{3}[64 - 8]$$

$$= \frac{56}{3}\pi\rho$$

So: $\qquad\qquad \bar{x} = \dfrac{56\pi\rho}{3(6\pi\rho)} = \dfrac{56}{18} = \dfrac{28}{9} = 3\dfrac{1}{9}$

The centre of mass is at a distance $3\frac{1}{9}$ from O on the axis of symmetry.

Centres of mass of surfaces of revolution

When part of a **curve** is rotated through $360°$ about a fixed line we obtain a surface known as a **surface of revolution**. Examples of such surfaces are a hollow cone and a hemispherical shell.

The centre of mass of a surface of revolution is in general difficult to obtain but we can obtain its position for a hemispherical shell by using calculus methods similar to those used for a volume of revolution.

The hemispherical shell

The hemispherical shell may be obtained by rotating the quarter of the circle $x^2 + y^2 = a^2$ in the 1st quadrant about the x-axis as shown in the figure.

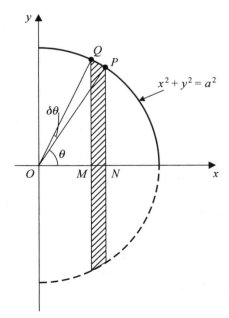

Consider the shaded strip obtained by rotating the arc PQ of the circle about the x-axis. Since the circle is of radius a the length of arc PQ is $a\delta\theta$, where $\angle POQ = \delta\theta$ and $\angle xOP = \theta$.

The shaded strip is approximately the curved surface of a cylinder of radius NP and height $a\delta\theta$. So the area is:

$$2\pi(NP)(a\delta\theta) = 2\pi a \sin\theta \, a\delta\theta$$

Therefore the mass of the shaded strip is:

$$2\pi\rho a^2 \sin\theta \, \delta\theta$$

where ρ is the mass per unit area.

Hence the total mass of the hemispherical shell, M, is:

$$M = 2\pi \int_0^{\frac{\pi}{2}} \rho a^2 \sin\theta \, d\theta = 2\pi\rho a^2 [-\cos\theta]_0^{\frac{\pi}{2}}$$
$$= 2\pi\rho a^2$$

[You could obtain the same result using the formula $4\pi a^2$ for a sphere, dividing by 2 to obtain the surface area of a hemisphere and then multiplying by ρ.]

Using $M\bar{x} = \sum_i m_i x_i$ with $m_i = 2\pi\rho a^2 \sin\theta \, d\theta$ and $x_i = ON = a\cos\theta$

then:
$$M\bar{x} = \int_0^{\frac{\pi}{2}} (2\pi\rho a^2 \sin\theta) \, d\theta \, a\cos\theta$$

So:
$$M\bar{x} = 2\pi\rho a^3 \int_0^{\frac{\pi}{2}} \sin\theta \cos\theta \, d\theta$$
$$= \pi\rho a^3 \int_0^{\frac{\pi}{2}} \sin 2\theta \, d\theta$$

since $\sin\theta \cos\theta = \frac{1}{2}\sin 2\theta$

Integrating:
$$M\bar{x} = \pi\rho a^3 \left[-\tfrac{1}{2}\cos 2\theta\right]_0^{\frac{\pi}{2}}$$
$$= \pi\rho a^3$$

So:
$$\bar{x} = \frac{\pi\rho a^3}{2\pi\rho a^2} = \frac{a}{2}$$

> Note: This result is given in the formula sheet. You will not be expected to reproduce this proof in your M3 exam.

Hollow cone

To obtain the centre of mass for a hollow cone a rather different method is used. Consider the surface of the cone to be made up of elementary strips each of which is a triangle with one vertex the vertex V of the cone as in the figure.

As shown in example 1 of this chapter, the centre of mass of each triangle, which can be considered as isosceles, is on the axis of symmetry and $\frac{2}{3}$ of the distance from the vertex V to the opposite side. So the centre of mass of the hollow cone is also two-thirds of the way from the vertex V to the base and by symmetry will be on the axis of the cone. It is more usual to say that for a hollow cone of height h the centre of mass is on the axis and $\frac{1}{3}h$ from the base.

Composite bodies

In chapter 2 of Book M2 a composite plane figure was considered as a combination of simple plane laminae, the positions of whose centres of mass were known. Composite bodies may be considered in the same way and this is illustrated in the following example.

Example 10

A uniform solid cone of height h and base radius a is joined to a uniform solid hemisphere of radius a so that the circular faces of the bodies coincide. The cone and the hemisphere have the same density. Find the distance of the centre of mass G of this composite body from the vertex of the cone.

Let X be the centre of the base of the cone and the plane face of the hemisphere. Then VX is an axis of symmetry of both the cone and the hemisphere and so will be an axis of symmetry of the composite body. The centre of mass G_1 of the cone lies on this line, the centre of mass G_2 of the hemisphere lies on this line and so the centre of mass of the composite body G will also lie on this line.

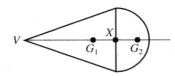

Recall that:

(i) the centre of mass of a uniform solid cone of height h is a distance $\dfrac{h}{4}$ from the base

(ii) the centre of mass of a uniform solid hemisphere of radius a is a distance $\dfrac{3a}{8}$ from the plane face.

So:
$$VG_1 = h - \frac{h}{4} = \frac{3h}{4}$$

$$VG_2 = VX + XG_2 = \left(h + \frac{3a}{8}\right)$$

Construct a table showing the masses of the separate bodies and the composite body, and the positions of the centres of mass as in chapter 2 of Book M2.

	Separate bodies		Composite body
	Cone	Hemisphere	
Mass	$\frac{1}{3}\pi a^2 h\rho$	$\frac{2}{3}\pi a^3 \rho$	$\frac{1}{3}\pi a^2 \rho(h+2a)$
Distance from V to centre of mass	$\dfrac{3h}{4}$	$\left(h+\dfrac{3a}{8}\right)$	$\bar{x}$

So:
$$M\bar{x} = \frac{1}{3}\pi a^2 \rho(h+2a)\bar{x} \qquad (1)$$

And:
$$M\bar{x} = (\tfrac{1}{3}\pi a^2 h\rho)\frac{3h}{4} + (\tfrac{2}{3}\pi a^3\rho)\left(h + \frac{3a}{8}\right)$$

$$= \tfrac{1}{4}\rho\pi a^2 h^2 + \tfrac{2}{3}\rho\pi a^3 h + \tfrac{1}{4}\rho\pi a^4$$

$$= \frac{\pi a^2\rho}{12}\left[3h^2 + 8ah + 3a^2\right] \qquad (2)$$

So from (1) and (2):
$$\bar{x} = \left(\frac{\pi a^2\rho}{12}\right)\left(\frac{3}{\pi a^2\rho}\right)\frac{[3h^2 + 8ah + 3a^2]}{(h + 2a)}$$

$$= \frac{1}{4}\left(\frac{1}{h + 2a}\right)[3h^2 + 8ah + 3a^2]$$

This is the distance of the centre of mass G of the composite body from the vertex of the cone.

Example 11

A uniform solid right circular cone has a base radius of 6 cm and a height of 12 cm. The top part of it, consisting of a small cone, base radius 3 cm and height 6 cm, is removed. Find the distance from the base of the large cone to the centre of mass of the remaining portion. (The remaining portion is called a **frustum**.)

A problem of this kind involving laminae was considered in chapter 2 of Book M2. The method used here is exactly the same in that the complete body is considered to be formed from the final frustum and the small cone which is removed.

Let O be the centre of the base of the large cone. Since OV is an axis of symmetry of both cones it will also be an axis of symmetry of the frustum. The centre of mass G of the frustum therefore lies on OV.

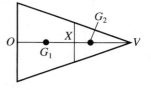

Let X be the point where OV meets the base of the small cone. Using the standard result for the centre of mass of a cone (see the table at the end of this chapter, or the formula book) then:

$$XG_2 = \tfrac{1}{4} \times 6\,\text{cm}$$

and $OG_2 = OX + XG_2 = 6\,\text{cm} + \tfrac{3}{2}\,\text{cm} = 7\tfrac{1}{2}\,\text{cm}$

$OG_1 = \tfrac{1}{4} \times 12\,\text{cm} = 3\,\text{cm}$

Also: mass of large cone $= \tfrac{1}{3}\pi(6)^2 12\rho = 144\pi\rho$

mass of small cone $= \tfrac{1}{3}\pi(3)^2 6\rho = 18\pi\rho$

So: mass of frustum $= 126\pi\rho$

	Separate bodies		Composite body
	Small cone	Frustum	Large cone
Mass	$18\pi\rho$	$126\pi\rho$	$144\pi\rho$
Ratio of masses	1	7	8
OG	$7\tfrac{1}{2}$ cm	$\bar{x}$	3 cm

So: $1 \times 7\frac{1}{2} + 7\bar{x} = 8 \times 3$

or: $7\bar{x} = 24 - 7\frac{1}{2} = 16\frac{1}{2}$

and $\bar{x} = 2\frac{5}{14}$ cm

The distance from the base of the larger cone to the centre of mass of the frustum is $2\frac{5}{14}$ cm.

Exercise 5B

You are advised to draw a sketch in these exercises.

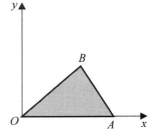

1   The vertices of the shaded triangular region shown in the figure are $O(0,0)$, $A(6,0)$ and $B(3,3)$. The region is rotated through $360°$ about the x-axis. By considering its symmetries write down the coordinates of the centre of mass of the solid of revolution formed.

In questions 2, 3 and 4 use symmetry considerations to obtain the answer.

2   The finite region bounded by the curve with equation $y = x^2 - 2x$ and the x-axis is rotated completely about the x-axis. Find the coordinates of the centre of mass of the solid of revolution formed.

3   The finite region bounded by the x-axis and the curve with equation $y = \sin x$ between $x = 0$ and $x = \pi$ is rotated completely about the x-axis. Find the coordinates of the centre of mass of the solid of revolution formed.

4   The finite region bounded by the y-axis and the curve with equation $y^2 - y = x$ is rotated completely about the y-axis. Find the coordinates of the centre of mass of the solid formed.

5   Find the coordinates of the centre of mass of the solid formed when the finite region bounded by the curve with equation $y = x^2$, the line $x = 1$ and the x-axis is rotated completely about the x-axis.

6   Find the coordinates of the centre of mass of the solid formed when the finite region bounded by the curve with equation $y = x^3$, the line $x = 1$ and the x-axis is rotated completely about the x-axis.

7   Find the coordinates of the centre of mass of the solid formed when the finite region bounded by the curve with equation $y = (1 + x^2)$, the lines $x = 0$ and $x = 1$ and the x-axis is rotated completely about the x-axis.

8   Find the coordinates of the centre of mass of the solid formed when the finite region bounded by the curve with equation $y = \dfrac{1}{x}$, the lines $x = 1$ and $x = 2$ and the x-axis is rotated completely about the x-axis.

9   Find the coordinates of the centre of mass of the solid formed when the finite region bounded by the curve with equation $y = \dfrac{1}{x^2}$, the lines $x = 1$ and $x = 2$ and the x-axis is rotated completely about the x-axis.

10   Find the coordinates of the centre of mass of the solid formed when the finite region bounded by the curve with equation $y = e^x$, the lines $x = 0$ and $x = 1$ and the x-axis is rotated completely about the x-axis.

11   Find the coordinates of the centre of mass of the solid formed when the finite region bounded by the curve with equation $y = e^{-x}$, the lines $x = 0$ and $x = 1$ and the x-axis is rotated completely about the x-axis.

12   A uniform cone C has base radius r and height h. It is joined to a uniform solid right circular cylinder, of the same density, with radius r and length l, so that the plane base of C coincides with a plane face of the cylinder. Find the distance from the vertex of C of the centre of mass of the composite body.

13   The base radius of the uniform solid cone C_1 is r and the radius of the uniform solid right circular cylinder C_2 is also r. The cylinder and the cone are made of the same material. A composite body is formed by joining the two so that the plane face of the cone and an end of the cylinder coincide. The centre of mass of the composite body is at the centre of the base of the cone. Find the ratio of the height of the cone to the length of the cylinder.

14 A uniform solid cylinder is of radius 4 cm and height 6 cm. A cylindrical hole of radius 2 cm and depth 3 cm is bored centrally at one end E. Find the distance of the centre of mass of the remainder from the end E.

15 A uniform solid cone C is of height h and base radius r. A frustum is cut from the cone by a plane parallel to the base at a distance $\frac{1}{2}h$ from the vertex. Show that the centre of mass of the frustum is at a distance $\frac{11}{56}h$ from its larger plane end.

16 The radii of the plane faces of the frustum of a cone are 2 cm and 3 cm and the distance between the faces is 5 cm. A uniform paper collar is fitted to the curved surface of the frustum. Find the distance of the centre of mass of the paper collar from the centre of the larger plane face of the frustum.

5.3 Simple cases of equilibrium of rigid bodies

In this section we consider situations in 3 dimensions where all the forces act in a vertical plane through the centre of mass of the body. Such problems are essentially two dimensional and we have a system of coplanar forces. The results used in chapter 2 of Book M2 for laminae can be generalised to this situation.

- **A rigid body is in equilibrium if**
  - **the vector sum of the forces acting is zero, that is the sum of the components of the forces in any given direction is zero.**
  - **the algebraic sum of the moments of the forces about any given point is zero.**

Suspension of a body from a fixed point

In chapter 2 of Book M2 it was shown that for a suspended lamina to be in equilibrium its centre of mass G must be vertically below the point of suspension.

The same result holds for a rigid body.

Consider a rigid body suspended from a point A. The only forces acting on the body are the force at A and the weight of the body. For the body to be in equilibrium these forces must be equal in magnitude and act in opposite directions along the same vertical line.

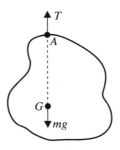

- **A rigid body hangs in equilibrium with its centre of mass vertically below the point of suspension.**

Example 12

A uniform solid right circular cone has base radius r and height $8r$. The cone is suspended by a string attached to a point A on the rim of its base. Find the angle between the axis of the cone and the downward vertical in the equilibrium position.

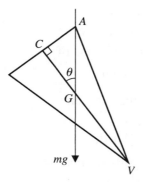

Using the standard formula $\dfrac{h}{4}$ for the distance of the centre of mass of a uniform solid cone from its base:

$$CG = \tfrac{1}{4}(8r) = 2r$$

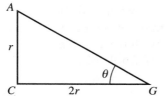

Let $\angle CGA$ be θ.

From the diagram: $\quad \tan \theta = \dfrac{r}{2r} = \tfrac{1}{2}$

So: $\qquad\qquad\qquad\qquad \theta = 26.6°$

The angle between the axis of the cone and the downward vertical is $26.6°$.

As in the case of a lamina (see chapter 2 of Book M2) it is often possible to find the required angle without redrawing the solid. Simply add G to your original diagram, join AG, and label this line as the vertical.

Equilibrium of bodies on a horizontal plane

Consider a body in contact with a horizontal plane. The figure shows a cross-section through the body which contains the centre of mass G. Suppose there is an area of contact represented in this cross-section by the line AB.

The weight of the body acts downwards through the centre of mass, and the only other vertical force acting on the body is the normal contact force or normal reaction R. If the body is to be in equilibrium in the position shown in the figure, the weight and R must be equal in magnitude and act in opposite directions along the same vertical line. Whether this is possible will depend on the position of G relative to the area of contact.

(i)   If the line of action of the weight lies **inside** the area of contact as in the figure above, then the body is in equilibrium and the force diagram is:

(ii) If the line of action of the weight lies **outside** the area of contact as shown here then it is not possible to have equilibrium as R must act somewhere in the area of contact. In this case the body will topple.

The intermediate case occurs when the line of action of the weight passes through either of the end points, A and B, of the area of contact. The body is then in equilibrium. This is the limiting case:

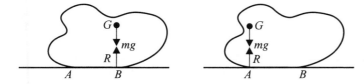

■ **For equilibrium the vertical through the centre of mass must fall in the area of contact.**

Example 13

A uniform square lamina $ABCD$ has side of length $4\,\text{cm}$. The point E lies on AD and is such that $ED = 3\,\text{cm}$. The triangle EDC is removed. Show that if the lamina is placed in a vertical plane with AE on a rough horizontal plane it will topple.

A mass m is to be placed at the corner B. Find the smallest value of m as a fraction of M, the mass of $ABCD$, that will prevent toppling.

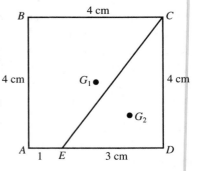

To calculate the distance of the centre of mass from AB proceed as in chapter 2 of Book M2.

The centre of mass of a square lies at the centre of the square and the centre of mass of a triangle is $\frac{2}{3}$ along a median from a vertex.

	Triangle	Remainder		Square
Mass	$\frac{1}{2} \times 3 \times 4\rho = 6\rho$	$16\rho - 6\rho = 10\rho$		$4 \times 4\rho = 16\rho$
Ratio of masses	3	5		8
Distance of centre of mass from AB	$4 - \frac{1}{3}(3) = 3$	$\bar{x}$		2
	(G_2)	(G)		(G_1)

where ρ is the mass per unit area.

Hence :
$$3 \times 3 + 5 \times \bar{x} = 8 \times 2$$
$$5\bar{x} = 16 - 9 = 7$$

So:
$$\bar{x} = 1.4\,\text{cm}$$

Since $AE = 1\,\text{cm}$ the line of action lies outside the area of contact and so the lamina will topple.

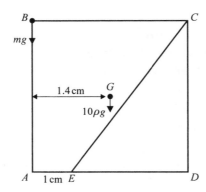

The body will turn about the point E when it topples. To prevent toppling the algebraic sum of the moments about E must be zero.

For lamina: moment about $E \curvearrowright (0.4)\,\text{cm} \times 10\rho g$

For mass: moment about $E \curvearrowleft (1)\,\text{cm} \times mg$

So: $(0.4)10\rho = m$

and: $m = 4\rho$

Since: $M = 4 \times 4\rho$

So: $\dfrac{m}{M} = \dfrac{4\rho}{4 \times 4\rho} = \dfrac{1}{4}$

Example 14

A uniform solid right circular cone has base radius 3 m and height 4 m. The top part of it, consisting of a small cone of base radius $1\frac{1}{2}$ m and height 2 m, is removed. Find the distance from the base of the larger cone to the centre of mass of the remaining portion and hence show that the frustum can rest in equilibrium with its curved surface in contact with a horizontal plane.

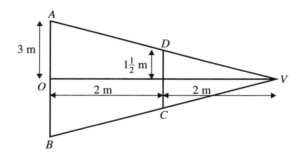

The calculation of the position of the centre of mass proceeds as before.

	Small cone	Frustum		Large cone
Mass	$\frac{1}{3}\pi\left(1\frac{1}{2}\right)^2 2\rho$	$12\pi\rho - 1\frac{1}{2}\pi\rho$		$\frac{1}{3}\pi(3)^2 4\rho$
	$= 1\frac{1}{2}\pi\rho$	$= 10\frac{1}{2}\pi\rho$		$= 12\pi\rho$
Ratio of masses	1	7		8
Distance of centre of mass from AB	$2 + \frac{1}{4}(2) = 2\frac{1}{2}$	$\bar{x}$		$\frac{1}{4}(4) = 1$

Hence: $2\frac{1}{2} + 7\bar{x} = 8$

or: $7\bar{x} = 8 - 2\frac{1}{2} = 5\frac{1}{2}$

So: $\bar{x} = \frac{11}{14}$ m

Consider now the situation with the curved surface in contact with a horizontal plane.

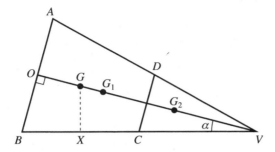

The triangle OBV is a right-angled triangle and since $OB = 3\,\text{m}$ and $OV = 4\,\text{m}$, so by Pythagoras' theorem $BV = 5\,\text{m}$. Also $CV = \frac{1}{2}BV = 2\frac{1}{2}\,\text{m}$. The distance GV of G, the centre of mass of the frustum, from V is $4\,\text{m} - \frac{11}{14}\,\text{m} = \frac{45}{14}\,\text{m}$

If the line of action of the weight cuts BV at X then:

$$VX = \tfrac{45}{14}\cos\alpha$$

From $\triangle OBV$, $\cos\alpha = \frac{4}{5}$ and so:

$$VX = \tfrac{45}{14}\left(\tfrac{4}{5}\right) = \tfrac{18}{7} = 2\tfrac{4}{7}\,\text{m} = 2.57\,\text{m}$$

Since $VX > VC = 2\frac{1}{2}\,\text{m}$ the line of action of the weight lies inside the area of contact. The frustum can therefore rest in equilibrium with its curved surface in contact with the horizontal plane.

Equilibrium of bodies on an inclined plane

Consider a body in contact with an inclined plane. The figure shows a cross-section through the plane and the body which contains the centre of mass G and a line of greatest slope of the plane.

Suppose there is an area of contact represented in this cross-section by the line AB. The forces acting on the body are: the weight of the body; the normal contact force R, perpendicular to the plane; the frictional force F parallel to the plane. For equilibrium, these three forces must all act through the same point *and* the resultant of R and F must be equal in magnitude to the weight of the body and act vertically upwards.

As in the previous discussion:

■ **if the line of action of the weight lies *inside* the area of contact the body is in equilibrium.**
If the line of action of the weight lies outside the area of contact then the body cannot rest in equilibrium and the body will topple.

Example 15

A solid uniform cylinder of radius r and length r has a solid uniform right circular cone, made from the same material, of base radius r and height r joined to it so that the base of the cone coincides with one end of the cylinder. The composite body is placed with its circular face on a rough plane which is inclined at an angle α to the horizontal. The plane is sufficiently rough to prevent sliding. Show that the body will topple over if $\tan \alpha > \frac{16}{11}$.

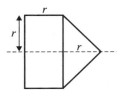

The calculation of the position of the centre of mass is shown in the usual tabular form:

	Cylinder	Cone	Composite body
Mass	$(\pi r^2)r\rho$	$\frac{1}{3}(\pi r^2)r\rho$	$\frac{4}{3}\pi r^3 \rho$
Ratio of masses	3	1	4
Distance of G from circular end	$\dfrac{r}{2}$	$r + \dfrac{r}{4} = \dfrac{5r}{4}$	$\bar{x}$

where ρ is mass per unit volume.

So:
$$3 \times \frac{r}{2} + 1 \times \frac{5r}{4} = 4 \times \bar{x}$$

$$\frac{11r}{4} = 4\bar{x}$$

And:
$$\bar{x} = \frac{11}{16}r$$

The centre of mass lies on the axis of symmetry.

The limiting case is shown in the figure:

$$\tan \alpha = \frac{r}{\left(\frac{11}{16}\right)r} = \frac{16}{11}$$

So for the limiting case:

The body will therefore topple if α exceeds this value, that is if $\tan \alpha > \frac{16}{11}$.

More general problems of equilibrium

In the next two examples more general problems of equilibrium are considered. These involve other forces in addition to the weight of the body and the reaction.

Example 16

The figure shows a hemisphere, of mass m and radius a, with O the centre of the plane face and P a point on the rim of the plane face. The hemisphere is suspended from a horizontal beam by two vertical inelastic strings fastened at O and P. Given that the tension in one string is 3 times the tension in the other string, find the two possible values of the tangent of the angle θ between OP and the horizontal.

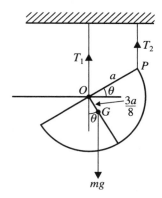

From the formula book the centre of mass G of the hemisphere is at a distance $\dfrac{3a}{8}$ from O on the axis of symmetry.

The diagram includes all the given information together with all the forces acting.

Let T_1 and T_2 be the tensions in the strings. Then either $T_1 = 3T_2$ or $T_2 = 3T_1$, giving two possible values for $\tan \theta$.

Resolving vertically: $\uparrow \ T_1 + T_2 - mg = 0$ (1)

Taking moments about $O: \curvearrowright mg\dfrac{3a}{8}\sin \theta = T_2 a \cos \theta$ (2)

(a) If $T_1 = 3T_2$ then from equation (1)
$$T_1 + T_2 = 4T_2 = mg$$
So:
$$T_2 = \tfrac{1}{4}mg$$
Substituting into equation (2):
$$\frac{3mga}{8}\sin\theta = \frac{mg}{4}a\cos\theta$$
So:
$$\frac{\sin\theta}{\cos\theta} = \frac{mga}{4}\times\frac{8}{3mga}$$
that is:
$$\tan\theta = \tfrac{2}{3}$$
(b) If $T_2 = 3T_1$ then from equation (1)
$$T_1 + T_2 = \tfrac{4}{3}T_2 = mg$$
So:
$$T_2 = \tfrac{3}{4}mg$$
Substituting into equation (2):
$$\frac{3mga}{8}\sin\theta = \frac{3mga}{4}\cos\theta$$
So:
$$\frac{\sin\theta}{\cos\theta} = \frac{3mga}{4}\times\frac{8}{3mga}$$
that is:
$$\tan\theta = 2$$

Example 17

A smooth sphere, mass 4 kg, radius 5 cm and centre O, rests against a smooth vertical wall. It is supported by a string of length 8 cm. One end of the string is attached to a fixed point A on the wall and the other end to a point on the surface of the sphere. Find:

(a) the inclination of the string to the wall

(b) the tension in the string

(c) the force exerted on the sphere by the wall.

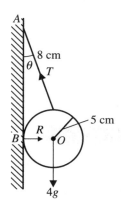

The diagram includes all the given information together with all the forces acting.

(a) Resolving horizontally: $\qquad \rightarrow R - T\sin\theta = 0 \qquad$ (1)

Resolving vertically: $\qquad \uparrow T\cos\theta - 4g = 0 \qquad$ (2)

Taking moments about B: $\qquad \circlearrowleft 4g \times 5 = TAB\sin\theta \qquad$ (3)

Eliminating T between equations (2) and (3) gives

$$\tan\theta = \frac{5}{AB} \qquad (4)$$

From equation (4) you can see that the line of action of T passes through O. (You could have obtained this result immediately by noting that the weight and R pass through O and so have zero moment about O. The tension T must also have zero moment about O as the three forces are in equilibrium. Hence the line of action of T passes through O.)

By Pythagoras: $\qquad AB^2 = 13^2 - 5^2 = 144$

So: $\qquad AB = 12 \,\text{cm}$

Hence: $\qquad \sin\theta = \frac{5}{13}$

$\qquad\qquad\qquad \cos\theta = \frac{12}{13}$

(b) From (2): $\qquad T \times \frac{12}{13} = 4g \quad \text{as } \cos\theta = \frac{12}{13}$

$$T = \frac{4 \times 13g}{12} = 42.5\,\text{N}$$

(c) From (1): $\qquad R = T\sin\theta$

$$= \frac{4 \times 13g}{12} \times \left(\tfrac{5}{13}\right) \quad \text{as } \sin\theta = \tfrac{5}{13}$$

$$= 16.3\,\text{N}$$

Exercise 5C

1 A uniform solid hemisphere is suspended by a string attached to a point on the rim of its base. Find the angle between the horizontal and the plane face of the hemisphere when it is in equilibrium.

2 A toy is constructed of thin uniform sheet metal and consists of a right circular cylinder with open ends of length l and radius r ($r < l$) covered by a hemispherical cap of radius r. Show that the centre of mass of the toy is at a distance $\frac{1}{2}(l+r)$ from the open end.

The toy is suspended from a point on the rim of the hemispherical cap. Show that, when the toy hangs in equilibrium, the axis of symmetry makes an angle α to the horizontal where:

$$\tan\alpha = \frac{l-r}{2r}$$

3 A square lamina $ABCD$ of side $4\,\text{cm}$ is made of uniform thin material. A semicircular piece, with AB as diameter, is removed. Show that the centre of mass of the remaining shape is at a distance:

$$\frac{40}{3(8-\pi)}\ \text{cm}$$

from AB.

The remaining shape is suspended by a string attached at C and hangs in equilibrium. Show that the angle θ between CD and the downward vertical is given by:

$$\tan\theta = \frac{2(14-3\pi)}{3(8-\pi)}$$

4 A uniform right circular cone of base radius a and height $3a$ rests in equilibrium with its base in contact with a rough inclined plane. The angle of inclination of the plane is then increased steadily. Assuming the plane is sufficiently rough to prevent slipping, find the angle of inclination of the plane to the horizontal when the cone is about to topple.

5 A uniform thin hemispherical bowl rests in equilibrium with its axis of symmetry horizontal and its curved surface on a rough inclined plane. Find the angle between the plane and the horizontal.

6 A uniform solid consists of a hemisphere of radius r and a right circular cone of base radius r and height h, fixed together so that their plane faces coincide. The solid can rest in equilibrium with any point of the curved surface of the hemisphere in contact with a horizontal plane. Find h in terms of r.

7 A uniform solid spinning top has the shape of an inverted right circular cone of radius $3r$ and height $4r$ surmounted by a right circular cylinder of base radius $3r$ and height $6r$. Find the position of the centre of mass of the spinning top and hence show that if it is placed with the curved surface of the cone on a horizontal plane the top will topple.

8 (a) Show, by integration, that the centre of mass of a uniform solid right circular cone of height h is at a distance $\frac{3}{4}h$ from the vertex.

Two uniform solid right circular cones, each of base radius a, have the same density and are of heights h and λh respectively where $\lambda > 1$. These cones are joined together, so that their circular bases coincide, to form a spindle as shown in the diagram.

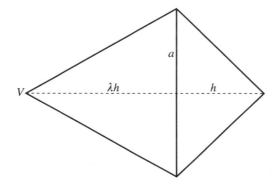

(b) Show that the centre of mass of this spindle is at a distance $\frac{1}{4}h(3\lambda + 1)$ from the vertex V of the larger cone.

Given that $a = h$:

(c) show that the spindle can rest in equilibrium with the curved surface of the smaller cone in contact with the horizontal plane provided that $\lambda \leqslant 5$.

9 A smooth plane is inclined at an angle of $\arcsin\left(\frac{3}{5}\right)$ to the horizontal. A solid sphere of mass $4\,\text{kg}$ and radius $7\,\text{cm}$ rests in contact with the plane. The sphere is maintained in equilibrium by a light inextensible string of length $18\,\text{cm}$ one end of which is attached to the sphere and the other to the plane.

(a) Show that the tension in the string is of magnitude $24.5\,\text{N}$.

(b) Show that the force exerted by the plane on the sphere is of magnitude $38.2\,\text{N}$.

10 A uniform solid right circular cone has base radius r, height $2r$
and mass m. One end of a light inextensible string is fastened
to the vertex of the cone and the other end is fastened to a
point on the rim of the base. When the string is hung over a
smooth peg, the axis of the cone is horizontal.
(a) Show that the two parts of the string are inclined at 45° to
the downward vertical.
(b) Find the tension in the string.

5.4 Slipping and toppling

When a force is applied to a body which is resting in equilibrium one
of several things may happen. It may remain at rest, it may slide or
it may tilt and topple over. Which of these occurs depends on
various factors, including the point of application of the force, the
magnitude of the force, the position of the centre of mass and the
coefficient of friction between the body and the surface on which it
is resting.

Consider a uniform cube of mass m and side $2a$ resting in
equilibrium on a rough horizontal plane. The diagram shows a
vertical section of this cube through its centre of mass G.

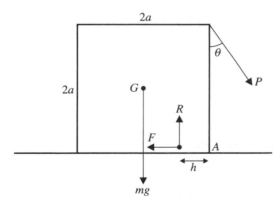

A force of magnitude P is applied to the mid-point of an upper edge,
as shown, at an angle θ to the downward vertical and in the plane of
the section.

In addition to P the forces acting are the weight of the cube and the
reaction of the plane, which is assumed to have components F and
R. Suppose this reaction acts at a distance h from A.

Resolving horizontally: $\rightarrow P\sin\theta - F = 0$ (1)

Resolving vertically: $\uparrow R - mg - P\cos\theta = 0$ (2)

Taking moments about A: $\circlearrowright Rh + P\,2a\sin\theta - mga = 0$ (3)

When $P = 0$, from equation (2) $R = Mg$ and so from equation (3) $h = a$ and the point at which the reaction acts is vertically below G, the centre of mass. As P increases, the point where the reaction acts moves to the right towards A and reaches A when

$$P\,2a \sin \theta - mga = 0 \qquad \text{from equation (3).}$$

That is:
$$P = \frac{mg}{2 \sin \theta}$$

If P increases beyond this value, equilibrium is not possible and the cube tilts.

So far we have assumed that the cube does not slide. But equilibrium can also be broken by sliding.

From equation (1): $\quad F = P \sin \theta$

From equation (2): $\quad R = mg + P \cos \theta$

As the cube is on the point of sliding, $F = \mu R$

so:
$$\mu = \frac{F}{R} = \frac{P \sin \theta}{mg + P \cos \theta}$$

Hence: $\quad \mu(mg + P \cos \theta) = P \sin \theta$

or: $\quad \mu mg = P(\sin \theta - \mu \cos \theta)$

so:
$$P = \frac{\mu mg}{(\sin \theta - \mu \cos \theta)}$$

Toppling will occur before sliding if the value of P which produces toppling is less than that which causes sliding, that is:

$$\frac{mg}{2 \sin \theta} < \frac{\mu mg}{(\sin \theta - \mu \cos \theta)}$$

so: $\quad \sin \theta - \mu \cos \theta < 2\mu \sin \theta$

hence: $\quad \sin \theta < \mu(\cos \theta + 2 \sin \theta)$

■ **To decide if equilibrium will be broken by sliding or toppling two cases need to be examined:**
   (i) when the body is on the point of sliding, so that $F = \mu R$,
   (ii) when the body is on the point of toppling, so that the reaction acts at the point about which the body will turn.

This is illustrated in the examples which follow.

Example 18

A box in the shape of a cube has mass 10 kg and side 1 m. It rests in equilibrium on a rough board, the coefficient of friction between the box and the board being 0.5. If the board is inclined at an angle α to the horizontal,
(a) find α, if the box is about to slide
(b) find α, if the box is about to topple.

If α increases from zero
(c) explain what happens to the box.

(a) Since the box is about to slide the forces acting are those shown:

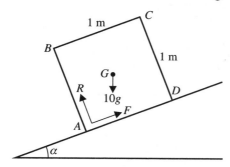

Resolving up plane: $\qquad\qquad\qquad\nearrow F - 10g \sin \alpha = 0$

Resolving perpendicular to plane: $\quad\searrow R - 10g \cos \alpha = 0$

When slipping occurs,
$$F = \mu R = \tfrac{1}{2} R$$
So:
$$10g \sin \alpha = \tfrac{1}{2} \times 10g \cos \alpha$$
$$\tan \alpha = \tfrac{1}{2}$$
$$\alpha = \arctan\left(\tfrac{1}{2}\right)$$
$$\alpha = 26.6°$$

(b) The box will topple about the horizontal axis through A. Since the only forces acting are the reaction at A and the weight of the box these must act along the same line.

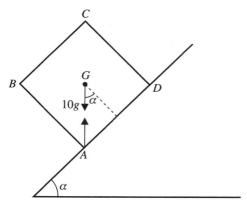

The centre of mass G must now be vertically above A.

Hence:
$$\tan \alpha = \frac{\tfrac{1}{2}}{\tfrac{1}{2}} = 1$$
$$\alpha = 45°$$

(c) If α is increased from zero the box will slip when $\alpha = 26.6°$ as this position is reached before $\alpha = 45°$, when toppling would occur.

Example 19

A uniform right circular cone has mass m, base radius r and height h. The cone rests in equilibrium on a rough horizontal plane, the coefficient of friction between the plane and the cone being μ. A horizontal force of magnitude P is applied at the vertex.

(a) Find the magnitude of P if the cone is on the point of slipping.
(b) Find the magnitude of P if the cone is on the point of toppling.

Given that the cone slides before it topples, as P increases from zero,
(c) obtain the range of possible values of μ.

(a) As the cone is on the point of slipping the forces acting are as shown in the figure.

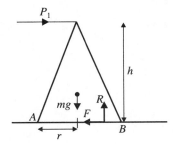

Resolving horizontally: $\rightarrow P_1 - F = 0$

Resolving vertically: $\uparrow R - mg = 0$

When slipping occurs $F = \mu R$.

So: $$P_1 = F = \mu R = \mu mg$$

(b) The cone will tilt about the point B. As the cone is on the point of toppling the reaction will act at B. The forces acting now are shown in the diagram below.

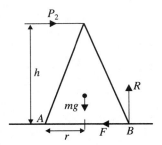

Taking moments about B: $\curvearrowright P_2 h - mgr = 0$

So: $$P_2 = mg\frac{r}{h}$$

(c) As the cone slides before it topples as P increases from zero, $P_1 < P_2$.

So:
$$\mu mg < mg\frac{r}{h}$$

$$0 < \mu < \frac{r}{h}$$

Example 20

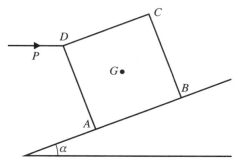

A uniform cube of mass 4 kg and side 1 m is placed on a rough plane of inclination α, where $\tan\alpha = \frac{1}{2}$. The coefficient of friction between the cube and the plane is $\frac{2}{3}$. The centre of mass G lies in the plane $ABCD$ and the edges perpendicular to this plane are horizontal.

A horizontal force P, steadily increasing in magnitude from zero, is applied at D (acting from left to right and with line of action lying in the plane $ABCD$). Show that equilibrium will be broken by the cube turning about the edge through B before it slides up the plane.

(i) Consider the situation when the cube is on the point of sliding. The relevant forces are shown in the diagram.

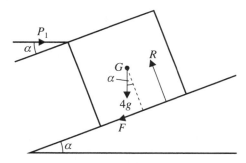

Resolving up the plane: $\nearrow\ P_1\cos\alpha - 4g\sin\alpha - F = 0$

Resolving normal to the plane: $\nwarrow\ R - 4g\cos\alpha - P_1\sin\alpha = 0$

$$\tan\alpha = \tfrac{1}{2}$$

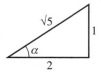

So: $\sin\alpha = \dfrac{1}{\sqrt{5}}, \quad \cos\alpha = \dfrac{2}{\sqrt{5}}$

For slipping: $F = \mu R = \tfrac{2}{3}R$

so: $P_1\dfrac{2}{\sqrt{5}} - 4g\dfrac{1}{\sqrt{5}} = \dfrac{2}{3}\left(4g\dfrac{2}{\sqrt{5}} + P_1\dfrac{1}{\sqrt{5}}\right)$

$$\frac{P_1}{\sqrt{5}}\left(2 - \tfrac{2}{3}\right) = \frac{4g}{\sqrt{5}}\left(\tfrac{4}{3} + 1\right)$$

$$\tfrac{4}{3} P_1 = 4 \times \tfrac{7}{3} \times g$$

$$P_1 = \frac{4 \times 7}{3}\left(\frac{3}{4}\right)g = 7g$$

(ii) Consider now the situation when the cube is on the point of toppling about the edge through B. The reaction will now act through B. The relevant forces are shown in the diagram

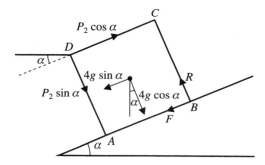

The easiest way to take moments about B is to resolve P_2 and the weight into components along and perpendicular to the plane:

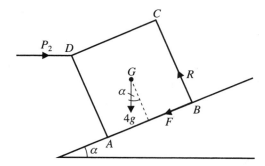

Taking moments about B:

$$\circlearrowright P_2 \cos\alpha \times 1 - P_2 \sin\alpha \times 1 - 4g\sin\alpha \times \tfrac{1}{2} - 4g\cos\alpha \times \tfrac{1}{2} = 0$$

or:

$$P_2\left(\frac{2}{\sqrt{5}} - \frac{1}{\sqrt{5}}\right) = 2g\left(\frac{2}{\sqrt{5}} + \frac{1}{\sqrt{5}}\right)$$

$$\frac{P_2}{\sqrt{5}} = 2g\frac{3}{\sqrt{5}}$$

$$P_2 = 6g$$

Since $P_2 < P_1$ equilibrium will be broken by the cube turning about the edge through B.

Exercise 5D

1 A uniform right circular cylinder of height h and radius r is placed with a plane face in contact with a rough horizontal plane, the coefficient of friction between the cylinder and the plane being μ. The plane is gradually tilted. Show that the cylinder will topple before it slides if $\mu > \dfrac{2r}{h}$.

2 A uniform right circular cone of height h and base radius a is placed with its base on a rough horizontal plane, the coefficient of friction between the cone and the plane being μ. The plane is gradually tilted and the cone begins to topple and slide at the same time. Show that $\mu = \dfrac{4a}{h}$.

3 In the following two problems a box in the shape of a cuboid is placed on a rough horizontal plane, the coefficient of friction between the box and the plane being 0.5. The plane is tilted gradually. The diagrams show a line of greatest slope and a cross section of the box. In each case:
(i) find the values of α at which the box would slide or topple
(ii) state which happens first.

(a)   (b)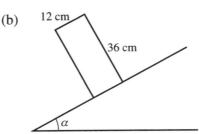

4 A uniform right circular cone of height h cm has base radius 20 cm. It is placed with its base on a rough plane inclined at 40° to the horizontal. Given that the cone is on the point of toppling find h.

5 A uniform solid cylinder of base radius r cm has height 20 cm. It is placed with one plane face on a rough plane inclined at 35° to the horizontal. Given that the cylinder is on the point of toppling, find r.

6

The diagram shows the central cross section of a uniform cube of mass 150 kg and side 1 m. It is placed on a rough horizontal plane, the coefficient of friction between the cube and the plane being 0.4. A force of magnitude P acts as shown at a corner parallel to the plane.

(a) Show that when $P = 735$ N the cube is about to topple.

(b) Find P if the cube is about to slide.

(c) If P is gradually increased from zero, does the cube slide or topple first?

7

The diagram shows a square lamina, of side a and mass m, standing upright on a rough horizontal plane, the coefficient of friction between the lamina and the plane being μ ($\mu < 1$).
'A string in the vertical plane of the lamina is attached to one corner and makes an angle of 45° with the downward vertical. A force P of gradually increasing magnitude acts along the string.

(a) Find the magnitude of P when the lamina is on the point of sliding.

(b) Find the magnitude of P when the lamina is on the point of tilting.

(c) Given that tilting occurs before sliding obtain the condition satisfied by μ.

8 A uniform wooden 'mushroom', used in a game, is made by joining a solid cylinder to a solid hemisphere. They are joined symmetrically, such that the centre O of the plane face of the hemisphere coincides with the centre of one of the ends of the cylinder. The diagram shows the cross section through a plane of symmetry of the mushroom, as it stands on a horizontal table. The radius of the cylinder is r, the radius of the hemisphere is $3r$, and the centre of mass of the mushroom is at the point O.

(a) Show that the height of the cylinder is $r\sqrt{\left(\frac{81}{2}\right)}$.

The table top, which is rough enough to prevent the mushroom from sliding, is slowly tilted until the mushroom is about to topple.

(b) Find, to the nearest degree, the angle with the horizontal through which the table top has been tilted.   [E]

SUMMARY OF KEY POINTS

1

The coordinates $(\bar{x}, \bar{y})$ of the centre of mass of the shaded region are given by:

$$\left[\int_a^b f(x)\,dx\right]\bar{x} = \int_a^b x f(x)\,dx$$

$$\left[\int_a^b f(x)\,dx\right]\bar{y} = \frac{1}{2}\int_a^b [f(x)]^2\,dx$$

2 The centre of mass of a solid body is the point at which the weight acts.

The weight of a uniform solid body is evenly distributed throughout its volume.

The centre of mass of a uniform solid body must lie on any **axis** of symmetry.

The centre of mass of a uniform solid body must lie on any **plane** of symmetry.

The centre of mass of a system of n masses $m_1, m_2, ..., m_n$ at points with coordinates (x_1, y_1, z_1), (x_2, y_2, z_2), ..., (x_n, y_n, z_n) has coordinates $(\bar{x}, \bar{y}, \bar{z})$ where

$$\bar{x} = \frac{\sum m_i x_i}{\sum m_i}$$

$$\bar{y} = \frac{\sum m_i y_i}{\sum m_i}$$

$$\bar{z} = \frac{\sum m_i z_i}{\sum m_i}$$

3

The x-coordinate of the centre of mass, $\bar{x}$, of the solid of revolution generated when the shaded region is rotated through 360° about the x-axis is given by:

$$M\bar{x} = \left[\int_a^b \rho\pi y^2 \mathrm{d}x\right]\bar{x} = \int_a^b \rho\pi y^2 x\,\mathrm{d}x$$

4 Standard results for uniform bodies

Body	Centre of mass
Solid hemisphere, radius r	$\frac{3}{8}r$ from centre
Hemispherical shell, radius r	$\frac{1}{2}r$ from centre
Circular arc, radius r, angle at centre 2α	$\dfrac{r\sin\alpha}{\alpha}$ from centre
Sector of circle, radius r, angle at centre 2α	$\dfrac{2r\sin\alpha}{3\alpha}$ from centre
Solid right circular cone, height h	$\dfrac{3h}{4}$ from vertex
Conical shell, height h	$\dfrac{2\lambda}{3}$ from vertex

5 A rigid body is in **equilibrium** if:
(i)   The vector sum of the forces acting is zero, that is the sum of the components of the forces in any given direction is zero.
(ii) the algebraic sum of the moments of the forces about any given point is zero.

6 A rigid body hangs in equilibrium with its centre of mass vertically below the point of suspension.
For a body in contact with a horizontal or inclined plane to be in equilibrium, the line of action of the weight, through the centre of mass, must lie inside the area of contact.

7 To decide if equilibrium will be broken by sliding or toppling two cases need to be examined:
(i)   when the body is on the point of sliding so that
$F = \mu R$
(ii) when the body is on the point of toppling, so that the reaction acts at the point about which the body will turn.

Review exercise 2

Whenever a numerical value of g is required take $g = 9.8\,\mathrm{m\,s^{-2}}$.

1 A particle of mass $0.4\,\mathrm{kg}$ is tied to one end of a taut light inextensible string of length $0.5\,\mathrm{m}$. The other end of the string is tied to a fixed point A on a smooth horizontal plane. Given that the particle moves with speed $8\,\mathrm{m\,s^{-1}}$ on the plane and in a circular path with centre A, find, in N, the tension in the string. Given that the string snaps when its tension exceeds $60\,\mathrm{N}$, find, in radians per second to one decimal place, the greatest angular speed at which the string can rotate.   [E]

2 A particle P, moving on the smooth inside surface of a fixed spherical bowl, of radius r, describes a horizontal circle at depth $r/2$ below the centre of the bowl. Show that P takes time $\pi\sqrt{(2r/g)}$ to complete one revolution of its circular path.   [E]

3 One end of a light inextensible string, of length a, is tied to a fixed point O and a particle Q, of mass m, is tied at the other end. The particle moves with constant angular speed ω in a horizontal circle so that the string traces out the surface of a cone. The centre of the circle is C, where C is vertically below O and angle COQ is β.
   (a) Find, in terms of m, g and β, the magnitude of the tension in the string.
   (b) Find ω^2 in terms of a, g and β, and hence deduce that
$$a\omega^2 > g \qquad\qquad [E]$$

4 A light inextensible string, of length $2\,\mathrm{m}$, passes through a small fixed smooth ring R and carries particle A, of mass $2M$, at one end and particle B, of mass M, at the other end. The particle A is at rest vertically below R. The particle B moves in

a horizontal circle, the centre of which is vertically below R, with constant angular speed $4\,\text{rad}\,\text{s}^{-1}$. Given that the string remains taut, calculate the distance AR.   [E]

5 A particle is attached at one end of a light string, the other end of which is fixed. When the particle moves in a horizontal circle with speed $2\,\text{m}\,\text{s}^{-1}$, the string makes an angle arctan $(5/12)$ with the vertical. Show that the length of the string is approximately $2.5\,\text{m}$.   [E]

6 An elastic string of length l and modulus $3mg$ has one end fixed to a point O. A particle of mass m is attached to the other end of the string and hangs in equilibrium. Find the stretched length of the string.
The particle is set in motion and describes a horizontal circle with constant angular speed ω. Given that the centre of the circle is at a distance l vertically below O, show
(a) that the stretched length of the string is $3l/2$
(b) that $\omega^2 l = g$.   [E]

7 A particle, of mass m, is suspended from a fixed point O by a light elastic string, of natural length l and modulus λ. The particle moves with constant angular speed ω in a horizontal circular path with the string making a constant angle θ with the downward direction of the vertical. Show that x, the extension of the string, is given by

$$\lambda x = \omega^2 lm(l + x)$$

Deduce that the motion described cannot take place unless $\omega^2 < \lambda/(lm)$.
Show further that, for a given value of ω, the depth of the horizontal circle below O is independent of λ.   [E]

8 (i) A bead, of mass m, threaded on a fixed, smooth, horizontal circular hoop, of radius d, is moving with constant speed v. Given that the force exerted by the hoop on the bead is of magnitude R, and that this force acts at an angle θ to the horizontal, express
(a) R^2 in terms of v, d, m and g
(b) $\tan\theta$ in terms of v, d and g.   [E]

(ii) State the model you used for the bead in this question. The bead is said to be 'moving with constant speed v'. For this to happen a physical force in addition to the frictional force has to be assumed negligible. Describe this force in words.

9 A parcel is placed on the rough horizontal floor of a van. The coefficient of friction between the parcel and the floor is $\frac{3}{8}$. The van rounds a horizontal circular bend of radius 100 m. Find the maximum constant speed, in metres per second, to 3 significant figures, at which the van may travel in order that the parcel shall not slide. [E]

10 A light elastic string has natural length l and modulus of elasticity $2mg$. One end of the string is attached to a fixed point A. A particle P, of mass m, is attached to the other end of the string.
(a) When P hangs vertically below A in equilibrium, find the length of the string in terms of l.
The point C is at distance l vertically below A. The particle P moves at constant speed in a horizontal circle, centre C. The string AP is inclined at an acute angle θ to AC.
(b) Show that $\theta = 60°$.
(c) Find the tension in the string, giving your answer in terms of m and g.
(d) Find the time taken for P to move once around the circle, giving your answer in terms of π, l and g. [E]

11 A particle of mass m is suspended from a fixed point A by a light inextensible string of length l. The particle moves in a horizontal circle, whose centre is vertically below A, with constant angular speed ω and with the string taut and inclined at an angle θ to the downward vertical through A.
(a) Show that $\omega^2 l \cos \theta = g$.
The string is replaced by a light elastic string of natural length a and modulus of elasticity λ. The particle now moves in a horizontal circle, whose centre is vertically below A, with constant angular speed 2ω. The elastic string makes the same angle θ with the downward vertical.
(b) Show that $\dfrac{1}{4a} - \dfrac{1}{4l} = \dfrac{m\omega^2}{\lambda}$. [E]

12 An artificial satellite of mass m moves under the action of a gravitational force which is directed towards the centre, O, of the earth and is of magnitude F. The orbit of the satellite is a circle of radius a and centre O. Obtain an expression for T, the period of the satellite, in terms of m, a and F.

Show that, if the gravitational force acting on a body of mass m at a distance r from O is $m\mu/r^2$, where μ is a constant, then $T^2\mu = 4\pi^2 a^3$.

Assuming that the radius of the earth is 6400 km and that the acceleration due to gravity at the surface of the earth is $10\,\text{m\,s}^{-2}$, show that $\mu = (6.4)^2 10^{13}\,\text{m}^3\,\text{s}^{-2}$.

Hence, or otherwise, find the period of revolution, in hours to 2 decimal places, of the satellite when it travels in a circular orbit 600 km above the surface of the earth. [E]

13

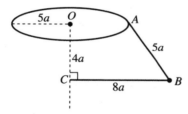

Two light inextensible strings AB and CB, of lengths $5a$ and $8a$ respectively, are each attached at B to a particle of mass m. The end A is attached to a point on the circumference of a circular disc of radius $5a$ and centre O, and the end C is attached to a point vertically below O such that $OC = 4a$. The disc, which remains in a horizontal plane, is made to rotate about the vertical axis OC, so that B moves in a horizontal circle, centre C, with constant angular speed ω. At each instant, both strings are taut and lie in a vertical plane through O, as shown in the diagram.

(a) Find, in terms of m and g, the tension in the string AB.

(b) Find, in terms of m, g, a and ω, the tension in the string BC.

(c) Hence show that

$$\omega^2 > \frac{3g}{32a}$$

[E]

14

The conical pendulum shown consists of a light inextensible string which has one end attached to a fixed point A. A particle P, of mass m, is attached to the other end of the string. The particle P moves with constant speed completing 2 orbits of its circular path every second and the tension in the string is $2mg$. Find, to the nearest cm,

(a) the radius of the circular path of P

(b) the length of the string. [E]

15 A child of mass 20 kg is seated on a swing, the ropes of which are 2 m long. She is pulled back by her father through an angle of 35° and pushed forwards with a speed of $2\,\mathrm{m\,s^{-1}}$. It is required to find her speed at the lowest point and the height she reaches, above this lowest point, on the other side.

It is suggested that the problem can be modelled by a particle of mass 20 kg attached to the end of a light inextensible string.

(a) Solve this model to find the required speed and distance.

(b) State the assumptions made using this model.

(c) Suggest ways in which this model can be refined.

16 A smooth circular hoop, centre O and radius a, is fixed with its plane vertical. A small smooth bead P, of mass $2m$, is threaded on the hoop. The bead is released from rest at a point on the same horizontal level as O. Show that if θ, $0 \leqslant \theta < \pi/2$, is the angle OP has turned through in time t

$$a\left(\frac{d\theta}{dt}\right)^2 = 2g\sin\theta.$$

When the bead reaches the lowest point A of the hoop it collides and coalesces with a small smooth bead of mass m, which is also threaded on the hoop and is at rest. Find the height, above A, to which the beads rise before coming to instantaneous rest.

Find also the magnitude of the force exerted by the beads on the hoop when the beads first return to A. [E]

17 A particle of mass m is suspended from a fixed point O by a
    light inextensible string of length a. When the string is hanging
    vertically the particle is projected horizontally with speed V.
    Find an expression for the tension in the string when it is
    inclined at an angle θ to the downward vertical.
    Given that $V^2 = \frac{7}{2}ag$, show that the string becomes slack
    when $\theta = \frac{2}{3}\pi$ and find the greatest height reached by the particle
    above the horizontal plane through the point of projection.   [E]

18 A particle of mass m is attached to one end A of a light
    inelastic string of length l. The other end of the string is
    attached to a fixed point B and the particle hangs in
    equilibrium under gravity. A horizontal impulse of magnitude
    J is applied to the particle.
    (a) Find the least possible value of J so that the particle can
    subsequently describe a complete vertical circle.
    (b) Given that $J = m\sqrt{(3lg)}$, find the tension in the string
    when it first becomes horizontal.
    Find also, in case (b), the height above the level of B to which
    the particle rises before the string first becomes slack.   [E]

19 A particle P is projected horizontally with speed u, where
    $u^2 < ag$, from the highest point A of a fixed smooth sphere of
    radius a and whose centre is at O. The particle slides on the
    outer surface of the sphere. Show that it leaves the sphere
    when PO makes an angle $\arccos\left(\frac{2}{3} + \frac{u^2}{3ag}\right)$ with the vertical.
    At this instant the speed is three times the initial speed. Show
    that $u = \sqrt{(ag/13)}$.   [E]

20 A small bead of mass m slides on a smooth circular hoop, with
    centre O and radius a, which is fixed in a vertical plane. The
    bead is projected with speed u, where $2ag < u^2 < 4ag$, from A,
    the lowest point of the hoop, and at a subsequent time t the
    bead is at a point P, where $\angle POA = \theta$. Find in terms of u, a, g
    and θ, an expression for $\left(\dfrac{d\theta}{dt}\right)^2$ and deduce that R, the
    magnitude of the force exerted by the hoop on the bead, is
    given by

$$aR = \mid mu^2 - mga(2 - 3\cos\theta)\mid$$

The bead is instantaneously at rest at the point B and the value of R is instantaneously zero at the point C. Show that the ratio of the vertical heights of B and C above the horizontal diameter of the hoop is $3:2$. [E]

21 A particle of mass m hangs at rest at one end of a light inextensible string of length l, the other end being fixed at the point O. The particle is then projected horizontally with speed u, where $u > \sqrt{(2gl)}$. Find the magnitude of the tension in the string when the string makes an acute angle α with the downward vertical through O.

When the string first becomes horizontal it comes into contact with a thin horizontal bar which is perpendicular to the plane of the motion and is at a distance $2l/3$ from O. Show that the particle describes complete vertical circles about the bar, provided that $u^2 \geqslant 3gl$. [E]

22

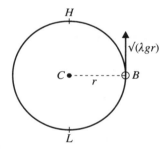

A bead B, of mass m, is threaded and can move freely on a smooth circular hoop, centre C and radius r. The hoop is fixed in a vertical plane. The bead B is projected vertically upwards with speed $\sqrt{(\lambda gr)}$ from a point level with C, as shown in the diagram. The highest point of the hoop is H and the lowest point is L.

(a) Show that, if the bead B passes through H, then $\lambda > 2$.

Given that $\lambda = 5$, find

(b) the speed of B at the instant when it passes through

    (i) H (ii) L

giving your answer in each case in terms of g and r

(c) the magnitude of the force exerted by B on the hoop at the instant when B passes through

    (i) H (ii) L

giving your answers in terms of m and g. [E]

23 A smooth sphere, with centre O and radius a, is fixed to the upper surface of a table. A smooth particle P is placed on the surface of the sphere, at a point A, where OA makes an angle β with the upward vertical $\left(0 < \beta < \dfrac{\pi}{2}\right)$. The particle is released from rest. Show that when OP makes an angle θ with the upward vertical, and the particle is still in contact with the sphere,

$$a\dot{\theta}^2 = 2g(\cos \beta - \cos \theta).$$

Determine the value of θ when the particle leaves the sphere. Given that $\cos \beta = \frac{3}{4}$, show that the particle hits the table at a distance $(\frac{3}{4}\sqrt{3})a$ from the vertical diameter. [E]

24

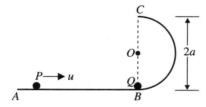

In a mechanics experiment, small marbles P and Q, of the same size but having masses $2m$ and m respectively, slide along a smooth groove in a rail.

The rail, which is fixed in a vertical plane, is in the form of a horizontal straight length AB joined to a semicircular arc BC of radius a and centre O, as shown.

Marble P is projected with speed u along AB towards marble Q, which is initially at rest at the point B. There is a direct collision between P and Q, immediately after which the speed of Q is $\frac{5}{4}u$.

(a) Find the speed of P immediately after the collision.

(b) Show that, for Q to remain in contact with the groove on the complete arc BC,

$$u \geqslant \sqrt{\left(\frac{16}{5}ga\right)}.$$ [E]

25 The diagram shows a finite region A which is bounded by the curve with equation $y^2 = 4ax$, the line $x = a$ and the x-axis. A uniform solid S_1 is formed by rotating A through 2π radians about the x-axis.

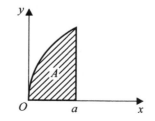

(a) Show that the volume of S_1 is $2\pi a^3$.

(b) Show that the centre of mass of S_1 is a distance $\dfrac{2a}{3}$ from the origin O. [E]

26 Show, by integration, that the centre of mass of a uniform solid hemisphere H, of base radius a, is at a distance $3a/8$ from the centre C of the plane circular base.

The plane face of a second hemisphere K, of radius $a/2$, and made of the same material as H, is stuck to the plane face of H, so that the centres of the two plane faces coincide at C, to form a uniform composite body S. Given that the mass of K is M, show that the mass of S is $9M$, and find, in terms of a, the distance of the centre of mass of the body S from C.

A particle P, of mass M, is attached to a point on the edge of the circular face of H of the body S. The body S with P attached is placed with a point of the curved surface of the part H in contact with a horizontal plane and rests in equilibrium. Find the tangent of the acute angle made by the line PC with the horizontal. [E]

27 Prove, by integration, that the position of the centre of mass of a uniform solid right circular cone is one quarter of the way up the axis from the base.

From a uniform solid right circular cone of height H is removed a cone with the same base and of height h, the two axes coinciding. Show that the centre of mass of the remaining solid S is a distance

$$\tfrac{1}{4}(3H - h)$$

from the vertex of the original cone.

The solid S is suspended by two vertical strings, one attached to the vertex and the other attached to a point on the bounding circular base. Given that S is in equilibrium, with its axis of symmetry horizontal, find, in terms of H and h, the ratio of the magnitude of the tension in the string attached to the vertex to that in the other string. [E]

28 Using integration, show that the centre of mass of a uniform
thin hemispherical bowl of radius a is at a distance $\frac{1}{2}a$ from
the centre C of the circular rim of the bowl.

A lid is attached to the bowl to form a closed composite body
B. The lid is a thin circular disc of radius a and centre C. The
lid is made of the same uniform material as the bowl. Show
that the centre of mass of B is at a distance $\frac{1}{3}a$ from C.

The body B has mass M. A particle P, also of mass M, is
attached at a point on the circumference of the plane circular
face of B. The body is placed with a point of its curved surface
in contact with a horizontal plane and rests in equilibrium.
Find, to the nearest half degree, the angle made by the line PC
with the horizontal.   [E]

29 A uniform semi-circular lamina has radius $2a$ and the
mid-point of the bounding diameter AB is O.

(a) Using integration, show that the centre of mass of the
lamina is at a distance $\dfrac{8a}{3\pi}$ from O.

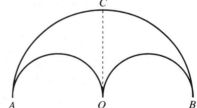

The two semi-circular laminas, each of radius a and with AO
and OB as diameters, are cut away from the original lamina to
leave the lamina $AOBC$ shown in the diagram, where OC is
perpendicular to AB.

(b) Show that the centre of mass of the lamina $AOBC$ is at a
distance $\dfrac{4a}{\pi}$ from O.

The lamina $AOBC$ is of mass M and a particle of mass M is
attached to the lamina at B to form a composite body.

(c) State the distance of the centre of mass of the body from
OC and from OB.

The body is smoothly hinged at A to a fixed point and rests in
equilibrium in a vertical plane.

(d) Calculate, to the nearest degree, the acute angle between
AB and the horizontal.   [E]

30 A uniform triangular lamina XYZ has $XY = XZ$ and the perpendicular distance of X from YZ is h. Prove, by integration, that the centre of mass of the lamina is at a distance $2h/3$ from X.

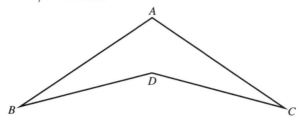

A uniform triangular lamina ABC has $AB = AC = 5a$, $BC = 8a$ and D is the centre of mass of the lamina. The triangle BCD is removed from the lamina, leaving the plate $ABDC$ shown in the diagram.

(a) Show that the distance of the centre of mass of the plate from A is $5a/3$.

The plate, which is of mass M, has a particle of mass M attached at B. The loaded plate is suspended from C and hangs in equilibrium.

(b) Prove that in this position CB makes an angle of $\arctan(1/9)$ with the vertical. [E]

31 A length of uniform thin wire is bent to form a semicircular arc of radius r. Use integration to show that the centre of mass is at a distance $\dfrac{2r}{\pi}$ from the diameter joining the end points.

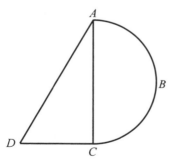

The diagram shows a plane frame made from uniform thin wire. The frame consists of straight wires AC, CD and DA and a semicircular wire ABC. The semicircular arc ABC is of radius 8 cm, angle $ACD = 90°$, and CD is of length $2x$ cm. Given that the centre of mass of the frame lies on AC, show that

$$3x^2 = 64$$

[E]

32 Use integration to show that the centre of mass of a uniform solid right circular cone, of vertical height h and base radius r, is at a distance $\frac{3}{4}h$ from the vertex V of the cone.

A uniform composite body consists of this cone and a solid hemisphere, of radius r, whose plane faces coincide. The centre of mass of the composite body is at G. The diagram shows a plane section of the body taken through V and G.

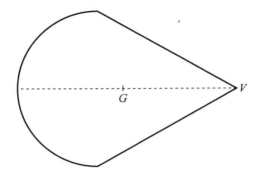

(a) Show that $VG = \dfrac{3h^2 + 8hr + 3r^2}{4(h + 2r)}$.

(b) If $VG = h$, find the semi-vertical angle of the cone.

(c) If $VG = r$, deduce that $3h = r(\sqrt{19} - 2)$.   [E]

33 A uniform triangular lamina ABC has $\angle ABC = 90°$ and $AB = c$. Using integration, show that the centre of mass of the lamina is at a distance $\frac{1}{3}c$ from BC.

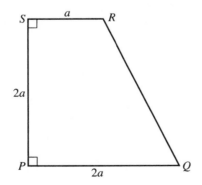

The diagram shows a uniform lamina in which

$$PQ = PS = 2a, SR = a, \angle SPQ = \angle RSP = 90°.$$

The centre of mass of this lamina is G.

(a) Show that the distance of G from PS is $\frac{7}{9}a$.

(b) Find the distance of G from PQ.   [E]

34 Use integration to show that the centre of mass of a uniform semi-circular lamina, of radius a, is at a distance $4a/3\pi$ from O the mid-point of its straight edge.

A semi-circular lamina, of radius b and with O as the mid-point of its straight edge, is removed from the first lamina. Show that the centre of mass of the resulting lamina is at a distance $\bar{x}$ from O, where

$$\bar{x} = \frac{4}{3\pi}\frac{(a^2 + ab + b^2)}{(a + b)}$$

Hence find the position of the centre of mass of a uniform semi-circular arc of radius a.   [E]

35

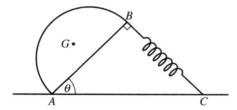

The diagram shows a uniform solid hemisphere of mass m, base radius a and with centre of mass G. The hemisphere is freely pivoted at a point A on the circumference of its plane face. The line AB is a diameter of the plane face and a light spring connects B to a point C on the horizontal plane through A. The system is in equilibrium with A, B, C and G in the same vertical plane, BC is perpendicular to AB and AB is inclined to AC at an angle θ, where $\tan\theta = 4/3$. Show that the thrust in the spring has magnitude $3mg/20$.

Given that the modulus of the spring is $3mg/4$, find the natural length of the spring.   [E]

36

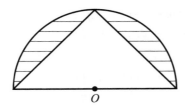

The diagram shows a cross-section of a solid formed by the removal of a right circular cone, of base radius a and height a, from a uniform solid hemisphere of base radius a. The plane bases of the cone and the hemisphere are coincident, both having centre O. Show that G, the centre of mass of the solid, is at a distance $a/2$ from O.

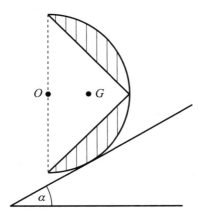

The second diagram shows a cross-section of the solid resting in equilibrium with a point of its curved surface in contact with a rough inclined plane of inclination α. Given that O and G are in the same vertical plane through a line of greatest slope of the inclined plane, and that OG is horizontal, show that $\alpha = \pi/6$.

Given that $\alpha = \pi/6$, find the smallest possible value of the coefficient of friction between the solid and the plane.   [E]

37 An experimental plastic traffic bollard B is made by joining a
uniform solid cylinder to a uniform solid right circular cone of
the same density. They are joined to form a symmetrical solid,
in such a way that the centre of the plane face of the cone
coincides with the centre of one of the plane faces of the
cylinder, as shown in the diagram.

The cylinder has radius $4r$ and height r. The cone has vertex
O, base radius $3r$ and height h.

(a) Show that the distance from O of the centre of mass of B is
$$\frac{32r^2 + 64rh + 9h^2}{4(16r + 3h)}$$

The bollard is placed on a rough plane which is inclined at an
angle α to the horizontal. The circular base of B is in contact
with the inclined plane. Given that $h = 4r$ and that B is on the
point of toppling,

(b) find α, to the nearest degree. [E]

38

A uniform solid cylinder, of base radius r and height h, has the
same density as a uniform solid hemisphere of radius r. The
plane face of the hemisphere is joined to a plane face of the
cylinder to form the composite solid S, shown in the diagram.
The point O is the centre of the plane base of S.

(a) Show that the distance from O of the centre of mass of S is
$$\frac{6h^2 + 8hr + 3r^2}{4(3h + 2r)}$$

The solid is placed on a rough plane which is inclined at an
angle α to the horizontal. The plane base of S is in contact
with the inclined plane. Given that $h = r$, and that S is on the
point of toppling,

(b) find α, to the nearest degree. [E]

39 A child's toy is made from joining a right circular uniform solid cone, radius r and height h, to a uniform solid hemisphere of the same material and radius r. They are joined so that their plane faces coincide, as shown in the diagram.

(a) Show that the distance of the centre of mass of the toy from the base of the cone is

$$\left| \frac{h^2 - 3r^2}{4(2r + h)} \right|$$

The toy is placed with its hemisphere in contact with a horizontal plane and with its axis vertical. It is slightly displaced and released from rest. Given that the plane is sufficiently rough to prevent slipping,

(b) explain clearly, with reasons, what will happen in each of the following cases:

    (i) $h > r\sqrt{3}$ (ii) $h < r\sqrt{3}$ (iii) $h = r\sqrt{3}$ [E]

40

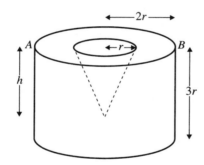

A mould for a right circular cone, base radius r and height h, is produced by making a conical hole in a uniform cylindrical block, base radius $2r$ and height $3r$. The axis of symmetry of the conical hole coincides with that of the cylinder, and AB is a diameter of the top of the cylinder, as shown in the diagram.

(a) Show that the distance from AB of the centre of mass of the mould is

$$\frac{216r^2 - h^2}{4(36r - h)}$$

The mould is suspended from the point A, and hangs freely in equilibrium.

(b) In the case $h = 2r$, calculate, to the nearest degree, the angle between AB and the downward vertical.   [E]

41

A uniform solid paperweight is in the shape of a frustrum of a cone, as shown. It is formed by removing a right circular cone of height h from a right circular cone of height $2h$ and base radius $2r$.

(a) Show that the centre of mass of the paperweight lies at a height of $\frac{11}{18}h$ from its base.

When placed with its curved surface on a horizontal plane, the paperweight is on the point of toppling.

(b) Find θ, the semi-vertical angle of the cone, to the nearest degree.   [E]

Examination style paper

M3

Answer all questions   **Time allowed 90 minutes**

Whenever a numerical value of g is required, take $g = 9.8 \, \text{m s}^{-2}$

1. A machine component consists of a small piston of mass $0.4 \, \text{kg}$ moving inside a fixed horizontal cylinder. The piston, which can be modelled as a particle, moves with simple harmonic motion performing 3 oscillations per minute. Given that the maximum speed of the piston is $0.5 \, \text{m s}^{-1}$,
    (*a*) find the amplitude of the motion,   **(4 marks)**
    (*b*) find the magnitude of the maximum force experienced by the piston.   **(3 marks)**

2. A light rod AB, of length $2 \, \text{m}$, is free to rotate in a vertical plane about a fixed horizontal axis through the end A. A particle of mass $0.5 \, \text{kg}$ is attached to the end B. When the rod is hanging at rest vertically downwards, a horizontal impulse of magnitude $5 \, \text{Ns}$ is applied to the particle at B. Show that the particle will complete vertical circles.   **(8 marks)**

3.

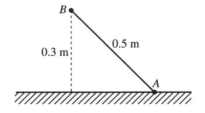

A stone, of mass $0.75 \, \text{kg}$, is attached to one end A of a light elastic string of natural length $0.4 \, \text{m}$ and modulus $12 \, \text{N}$. The other end B of the string is fixed to a point $0.3 \, \text{m}$ above rough horizontal ground. When $AB = 0.5 \, \text{m}$ the stone rests on the gound in limiting equilibrium as shown in the figure. Modelling the stone as a particle, calculate μ the coefficient of friction between the stone and the ground, giving your answer to 2 significant figures.   **(10 marks)**

4. A particle P moves in a straight line with acceleration $(6t)$ m s^{-2} at time t seconds. Its displacement from a fixed point O of the line at this time is x metres. Given that $x = 6$ when $t = 1$ and $x = 75$ when $t = 4$ find an expression for x in terms of t. **(11 marks)**

5.

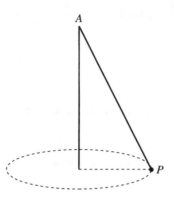

The conical pendulum shown in the above figure consists of a light inextensible string which has one end attached to a fixed point A. A particle P, of mass m, is attached to the other end of the string. The particle moves with a constant speed completing 2 orbits of its circular path every second. Given that the tension in the string is $4mg$ find, in cm to two significant figures,
(a) the radius of the circular path of P, **(9 marks)**
(b) the length of the string. **(2 marks)**

6. A light elastic string of natural length 2 m has modulus of elasticity $12g$ N. One end of the string is attached to a fixed point O and to the other end is attached a body of mass 3 kg. The body can be modelled as a particle. The body is initially held at O and then allowed to fall vertically.
(a) Use the conservation of energy principle to show that the body falls a distance of 4 m before coming to rest. **(6 marks)**
(b) Find the greatest speed of the body during the fall stating the distance from O at which it occurs. **(6 marks)**

7.

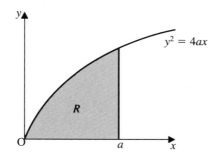

The shaded region R is bounded by the curve with equation $y^2 = 4ax\,(y > 0)$, the line $x = a$ and the x-axis. The region R is rotated completely about the x-axis to a form a solid S.

(a) Find the coordinates of the centre of mass of S.   **(7 marks)**

The solid S is joined to a uniform right circular cylinder, of the same density and base radius $2a$, so that the plane face of S coincides with a plane face of the cylinder. Given that the length of the cylinder is $4a$,

(b) find the distance of the centre of mass of the composite solid from its plane face.   **(6 marks)**

The composite solid is placed with its plane face in contact with a rough plane inclined at an angle α to the horizontal. The plane is sufficiently rough to prevent slipping.

(c) Show that the solid will topple over if $\tan \alpha > \dfrac{54}{61}$.   **(3 marks)**

Answers

Note on accuracy: Unless otherwise stated all numerical answers are given to 3 significant figures. When calculating answers the maximum number of figures possible should be used before the final answer is rounded to 3 significant figures. Rounding intermediate answers can produce an inaccurate final answer.

Edexcel accepts no responsibility whatsoever for the accuracy or method of working in the answers given for examination questions.

Exercise 1A

1 $21\,\mathrm{m\,s^{-1}}$ 2 $48\,\mathrm{m\,s^{-1}}$ 3 $9\,\mathrm{m\,s^{-1}}, 9\,\mathrm{m}$

4 $29\,\mathrm{m\,s^{-1}}, 24\,\mathrm{m}$ 5 $7\frac{1}{2}\mathrm{s}, 29\frac{1}{4}\mathrm{m}$

6 $19\,\mathrm{m\,s^{-1}}$

7 (a) $t = 1, 3$ (b) $4\,\mathrm{m}$

8 $1\,\mathrm{m\,s^{-1}}, \frac{\pi}{4}\mathrm{m}$

9 $t^3 + 2t + 2, 35$ 10 $x = \mathrm{e}^{-t} + 5t + 4$

11 $t = 5$ 12 $(-3\ln 7 + 18)\,\mathrm{m}$

Exercise 1B

1 $v^2 = u^2 + 2fx$ 2 $8\,\mathrm{m\,s^{-1}}$

3 $v^2 = 1600 - 40x - 2x^2, x = 20$

4 $\sqrt{\left(\dfrac{a^3 2\lambda}{3}\right)}$ 5 $x = \exp\left(\dfrac{v^2}{2k}\right)$ 6 2

7 (a) $v = x + 3$ (b) $x = 3\mathrm{e}^t - 3$

8 (a) $v = \dfrac{2}{x}$ (b) $x = \sqrt{4t + 1}$

9 (a) $v = -\mathrm{e}^x$ (b) $x = \ln\left(\dfrac{u}{ut + 1}\right)$

10 $t = \frac{1}{4}\sqrt{15}$ 11 $2\sqrt{13}\,\mathrm{m\,s^{-1}}$

12 $x = -\ln(1 - t)$

14 $1.29\,\mathrm{m\,s^{-1}}$

15 (b) $x = 2\mathrm{e}^t - 2$

Exercise 2A

1 $7.5\,\mathrm{N}$ 2 $0.0625\,\mathrm{m}$ 3 $0.563\,\mathrm{m}$

4 $8\,\mathrm{N}$ 5 $180\,\mathrm{N}$ 6 $3.13\,\mathrm{m}$

7 $12.8\,\mathrm{cm}$ 8 $0.94\,\mathrm{m}, 1.28\,\mathrm{m}$

9 $58.8\,\mathrm{N}$ 10 $1.96\,\mathrm{m}$ 11 $11.2\,\mathrm{N}, 8.4\,\mathrm{N}$

12 35 13 $6l$ 14 $19.8\,\mathrm{N}, 0.2\,\mathrm{N}$

15 Ball – dimensions small compared with length of cord, particle model. Cord –assume light, light elastic string model. $3.68\,\mathrm{N}, 61.3\,\mathrm{N}$

16 $35.2°, 13.9\,\mathrm{N}$

17 (a) $15.1\,\mathrm{N}$ (b) $0.603\,\mathrm{m}$

18 $2.04\,\mathrm{m\,s^{-2}}$

19 $8.49\,\mathrm{m\,s^{-2}}$

20 (a) Particle, light elastic spring, $44.6\,\mathrm{N}$
    (b) 0.172

21 (a) $12\,\mathrm{m\,s^{-2}}$ (b) $12\,\mathrm{m\,s^{-2}}$

23 $10\,\mathrm{m\,s^{-2}}$ upwards

Exercise 2B

1 $1\,\mathrm{J}$ 2 $1.5\,\mathrm{J}$

3 (a) $0.72\,\mathrm{J}$ (b) $4.62\,\mathrm{J}$

4 (a) $1.25\,\mathrm{J}$
    (b) $2.5\,\mathrm{J}$ They are different although change in length is the same.

5 $2.49\,\mathrm{J}$ 6 $\frac{1}{2}mgl$ 7 $\frac{3}{2}mgl$ 8 $\frac{1}{4}mgl$

Exercise 2C

1 $1.37\,\mathrm{m\,s^{-1}}$

2 (a) $1.73\,\mathrm{m\,s^{-1}}$ (b) $1.63\,\mathrm{m\,s^{-1}}$

3 (a) $3.34\,\mathrm{m}$ (b) $3.82\,\mathrm{m\,s^{-1}}$

4 $0.180\,\mathrm{m}$

7 (b) $4l$

8 $0.665\,\mathrm{m}, 1.25\,\mathrm{m\,s^{-1}}$

9 $48.1\,\mathrm{N}, 2.90\,\mathrm{m\,s^{-1}}$

10 Assume stone is small, model as particle; assume elastic light.

(b) $20.1\,\text{m s}^{-1}$

(c) Answer different: gravitational P.E. would have to be included.

Exercise 2D

1 (a) $1.65\,\text{m}$ (b) $5.56\,\text{m s}^{-1}$ (c) $2.34\,\text{m}$

2 (a) $30\,\text{N}$ (b) $1.2\,\text{J}$ (c) increase

3 (a) $17.6\,\text{J}$ (b) $3.6\,\text{kg}$

4 (a) $\dfrac{5mg}{2}$ (b) $1.13\,\text{m}$ (c) $18°$

5 (a) $\sqrt{\left(\dfrac{13lg}{20}\right)}$ (b) $\frac{2}{5}\sqrt{\left(\dfrac{13lg}{20}\right)}$ (c) $\dfrac{13l}{50}$

6 (a) $\frac{2}{3}$ (b) $\sqrt{(\frac{2}{3}gl)}$ (c) $\dfrac{3l}{2}$

7 (b) $8\,\text{m}$

Exercise 3A

1 (a) $3.68\,\text{m s}^{-1}$ (b) $15.9\,\text{m s}^{-1}$
(c) $t = 2\pi\,\text{s}$ (d) $22.9\,\text{m}$ (e) $16\pi\,\text{m}$

2 (a) $3.74\,\text{m s}^{-1}$ (b) $5.10\,\text{m s}^{-1}$

3 (a) $54\,\text{m s}^{-1}$ (b) $6.9\,\text{m}$

4 (a) $4\,\text{m s}^{-1}$ (b) $28\,\text{m}$
(c) $\frac{5}{3}\,\text{s}$ (d) $25.9\,\text{m}$

5 (a) $9.90\,\text{m s}^{-1}$ (b) $1.70\,\text{s}$

6 (a) $28\,\text{N s}$ (b) $0.549\,\text{N s}$
(c) $15.5\,\text{N s}$ (d) $0.6\,\text{N s}$

7 (a) $66.7\,\text{J}$ (b) $8.13\,\text{J}$ (c) $7.39\,\text{J}$

8 (a) $32\,\text{N s}$ (b) $32\,\text{m s}^{-1}$ (c) $6\,\text{s}$

9 $\dfrac{0.8}{3}(x^2 + x)$, $14.4\,\text{J}$

Exercise 3B

1 (a) $1.67 \times 10^{-8}\,\text{N}$ (b) $1.67 \times 10^{-12}\,\text{N}$
(c) $1.67 \times 10^{-18}\,\text{N}$

2 $0.534\,\text{kg}$

3 $9.78\,\text{m s}^{-2}$ Earth can be modelled as a particle

4 $5.96 \times 10^{24}\,\text{kg}$ 5 $1.46 \times 10^{2}\,\text{kg}$

6 $1.62\,\text{m s}^{-2}$ Particle model

7 (a) $1.49\,\text{km s}^{-1}$ (b) $1160\,\text{km}$

8 $2.37\,\text{km s}^{-1}$

Exercise 3C

1 $\dfrac{3\pi}{10}$, $1.89\,\text{m s}^{-1}$ 2 $2.5\,\text{m s}^{-1}$

3 $\frac{5}{6}\,\text{m}$, $1.81\,\text{m s}^{-1}$ 4 $2\pi\,\text{s}$, $10\,\text{m}$

5 $11.9\,\text{m s}^{-1}$ 6 $0.5\,\text{m s}^{-1}$, $0.433\,\text{m s}^{-1}$

7 $\frac{5}{2}\,\text{m}$, $x = \frac{5}{2}\sin 2t$ 8 $\dfrac{3}{4\pi}\,\text{m}$, $0.0475\,\text{m}$

9 $2.53\,\text{m}$, $1.22\,\text{s}$

10 $3.29\,\text{J}$ Piston moves in a straight line and can be modelled as a particle.

11 Particle model, $0.785\,\text{m s}^{-1}$, $0.912\,\text{s}$

12 $5\,\text{m}$, $\dfrac{2\pi}{3}\,\text{s}$ (a) $15\,\text{m s}^{-1}$ (b) $0.280\,\text{s}$ (c) $0.175\,\text{s}$

13 $2\,\text{m}$, $\dfrac{2\pi}{5}\,\text{s}$, $0.959\,\text{m}$, $0.528\,\text{s}$

14 $0.598\,\text{s}$ 15 $1.46\,\text{s}$

16 $2.07\,\text{pm}$ $10.33\,\text{pm}$

Exercise 3D

1 $0.726\,\text{s}$, $0.289\,\text{m}$

2 $x = 0.129\sin 12.9t$, $21.5\,\text{m s}^{-2}$

3 $0.444\,\text{s}$, $2.83\,\text{m s}^{-1}$

4 (a) Particle, light elastic string, smooth horizontal surface
(b) $3.79\,\text{m s}^{-1}$ (c) $1.55\,\text{s}$
(d) Smooth surface $\Rightarrow$ frictional force negligible. The answer to (b) would be smaller.

5 $0.5\,\text{m}$, $2.11\,\text{m s}^{-1}$, $0.25\,\text{m}$

6 (a) Spring light compared with piston; model spring as light elastic spring; assume piston can be modelled as a particle.
(b) $0.251\,\text{s}$, $2.5\,\text{m s}^{-1}$
(c) Cylinder unlikely to be perfectly smooth – include frictional force.

7 $1.22\,\text{m s}^{-1}$, $0.976\,\text{s}$

8 $0.5\,\text{m}$, $1.72\,\text{s}$, $1.83\,\text{m s}^{-1}$, $1.25\,\text{J}$

9 $2.08\,\text{m s}^{-1}$ $2.16\,\text{J}$

10 l, $2\pi\sqrt{\left(\dfrac{l}{10g}\right)}$, $\frac{1}{2}\sqrt{(30gl)}$

Exercise 3E

1 $0.49\,\text{m}$, $1.40\,\text{s}$, $2.24\,\text{m s}^{-1}$

2 $0.513\,\text{s}$, $0.163\,\text{m}$

3 $45\,\text{N}$, $0.8\,\text{m}$

4 (a) $1.25\,\text{m}$ (b) $0.25\,\text{m}$ (c) $1.5\,\text{m s}^{-1}$

5 Man – particle; spring – light elastic; assume mass of man large compared with spring; assume motion is in a straight line. $0.02\,\text{m}$, $0.284\,\text{s}$

6 Piston – particle; spring – light elastic; assume mass of piston large compared with that of spring, dimensions of piston small compared with length of spring.
0.196 s, 0.655 m s^{-1}
Include frictional force

7 0.295 s, 0.503 s

8 0.0143 m, 0.468 s

9 (a) Climber – particle; rope – light elastic string; assume rope long and light compared with height and weight of climber, initial speed of climber is 0 m s^{-1}
(b) 14 m s^{-1} (c) 13.0 m (d) 1.75 s

10 (a) 1.87 m (b) 0.1 m, $\pi\sqrt{\left(\frac{3}{20}\right)}$ s or 1.22 s

11 (d) $2\pi\sqrt{\dfrac{l}{g}}$

Exercise 3F

1 31.4 m s^{-1} **2** (a) 4 m (b) $1\frac{7}{8}$ J

3 (a) $\dfrac{8l}{3}$ (c) $2\pi\sqrt{\left(\dfrac{2l}{3g}\right)}$ (d) $\sqrt{\left(\dfrac{gl}{6}\right)}$

4 (a) $v^2 = 5x^2 - 61$ (b) $\sqrt{70}$ m

5 (a) 24.5 N (c) 0.550 s (d) 2.04 J

6 (b) 8 s, 6 m (c) $\dfrac{3\pi}{2}$ m s^{-1} (d) $\frac{4}{3}$ s

7 (a) 202 N (c) $\frac{1}{9}(e^{3t} - 3t - 1)$

8 $\sqrt{2\pi}$ s

Review exercise 1

1 53.2 m

2 $a = 6 + 2t$, $x = 3t^2 + \frac{1}{3}t^3 - \frac{44}{3}$, 102 m

3 20 m **4** (a) $1\frac{5}{27}$ m (b) 2 s (c) 4 m s^{-1}

5 (b) $43\frac{1}{3}$ m

7 (a) $v = 1 + 2\cos 2t$ (b) $f = -4\sin 2t$

8 (b) 115 m (c) 130 J (d) 8 N

9 (b) $v = 9.8(e^t - 1)$
(c) Assumed small enough to be modelled as a particle.

10 $\sqrt{\left(\dfrac{3gh}{2}\right)}$

11 (a) 3.50 s (b) $\dfrac{19.6}{k}(1 - e^{-\frac{kt}{2}})$

12 (a) 21 N s (b) 47 m s^{-1}

13 (a) 12 J (b) 7.75 m s^{-1}

15 (a) $v\dfrac{dv}{dx} = x^2 + 4$ (c) 55.8 J

18 (a) $\sqrt{\left(\frac{2}{3}kc^3\right)}$ (b) $\frac{1}{3}mkc^3$

19 (b) $\dfrac{\lambda m}{3c}$ **20** $l + \dfrac{mgl}{\lambda}$

21 250 mm   **22** 0.3 m

23 118 N (a) 25.5 N (b) 14.7 N (c) 1.35 m

24 $\dfrac{79c}{36}$ **25** (a) 19.6 N (b) 43.1 N

26 12 : 5, 11.76 N, 106 N   **27** (a) $\dfrac{l}{4}$

28 (a) $5Mg$ (b) $20Mg$

29 (b) 4.8 N (c) 13.2 N

30 (b) $\frac{1}{3}\sqrt{(ga)}$

31 (a) 105 N (b) 70 N (c) 42 J

32 (b) $\dfrac{2mg}{l}(x^2 + lx)$
(c) $\frac{1}{2}mgl + 2mgx - mv^2$
(d) $v = \sqrt{\left(\dfrac{3gl}{8}\right)}$

33 (c) $e = \sqrt{\left(\dfrac{2a}{2a + b}\right)}$

34 (b) 6.64 m s^{-1} (c) 4 m

35 (a) $3l = \lambda(0.6 - l)$ (b) $\lambda = 15$

37 (a) $u = 15.4$ m s^{-1} **38** $x = a\sin nt$

39 (a) $-an\sin nt$, $-an^2\cos nt$ (b) a^2n^2
(c) $an^3\sin nt = -n^2v$, S.H.M.

40 $\dfrac{\sqrt{3}}{2}$ m s^{-1} **41** (a) $-k\omega\pi$ (b) $\dfrac{k\pi}{4\omega}$

42 (a) 2.5 m (b) $\dfrac{\pi}{4}$ s (c) -10 m s^{-1}
(d) 20 m s^{-2} (e) $\dfrac{2\pi}{3}$ s

43 $an\cos nt$, $-an^2\sin nt = -n^2s$, $a = 0.15$, $n = 4$, 0.785 s

44 (a) $\pi\sqrt{\left(\dfrac{2l}{g}\right)}$, $\dfrac{l}{4}$

45 (a) 0945 hours
(b) Particle moving in a straight line with S.H.M.

46 $2\pi\sqrt{\left(\dfrac{e}{g}\right)}$ **47** 3, $\dfrac{3l}{4}$

48 $2a\sqrt{\left(\dfrac{g}{l}\right)}$ **49** $\dfrac{l}{10}$, $2\pi\sqrt{\left(\dfrac{l}{5g}\right)}$

50 (a) $-\dfrac{3}{4}\pi^2\cos\dfrac{\pi}{2}t\,\mathrm{m\,s^{-2}}$ (b) $4\,\mathrm{s}$

    (c) $\dfrac{3\pi}{2}\,\mathrm{m\,s^{-1}}$ (d) $\dfrac{1}{3}\,\mathrm{s}$

51 (a) $\dfrac{3L}{2}$ (c) $-\dfrac{L}{2}\leqslant h\leqslant\dfrac{L}{2}$

    (d) $\pi\sqrt{\left(\dfrac{L}{2g}\right)}$ (e) $\dfrac{1}{3}\sqrt{(2gL)}$

Exercise 4A

1 $28.6\,\mathrm{rev\,min^{-1}}$ **2** $0.419\,\mathrm{rad\,s^{-1}}$
3 (a) $10\,\mathrm{rev\,min^{-1}}$ (b) $1.05\,\mathrm{rad\,s^{-1}}$
4 (a) $7.5\,\mathrm{rev\,min^{-1}}$ (b) $0.785\,\mathrm{rad\,s^{-1}}$
5 $12.5\,\mathrm{m\,s^{-1}}$ **6** $0.0785\,\mathrm{m\,s^{-1}}$
7 $41.9\,\mathrm{m\,s^{-1}}$ **8** $30\,\mathrm{rad\,s^{-1}}$, $1.26\,\mathrm{s}$
9 $1.8\,\mathrm{rad\,s^{-1}}$ **10** $0.4\,\mathrm{rad\,s^{-1}}$
11 (a) $1.6\,\mathrm{rad\,s^{-1}}$ (b) $1.2\,\mathrm{m\,s^{-1}}$

Exercise 4B

1 $0.206\,\mathrm{m\,s^{-2}}$ **2** $200\,\mathrm{m\,s^{-2}}$
3 (a) $10\,\mathrm{rad\,s^{-1}}$ (b) $15\,\mathrm{m\,s^{-1}}$
4 $5.6\,\mathrm{m\,s^{-2}}$ **5** $1.48\,\mathrm{m\,s^{-2}}$ **6** $0.0337\,\mathrm{m\,s^{-2}}$

Exercise 4C

1 $83.3\,\mathrm{N}$ **2** $2.5\,\mathrm{m\,s^{-1}}$ **3** $3.30\,\mathrm{rad\,s^{-1}}$
4 0.638 **5** $7.2\,\mathrm{N}$ **6** $2.5\,\mathrm{m}$
7 (a) $10\,\mathrm{N}$ (b) $17.3\,\mathrm{rad\,s^{-1}}$ **9** $13:10$
11 $3.91\,\mathrm{rad\,s^{-1}}$
12 $46.7\,\mathrm{N}$ at $9.66°$ to the horizontal.

Exercise 4D

1 Particle model, $2mg$, $\frac{1}{2}\sqrt{(6gr)}$
2 Particle, smooth surface, $3.13\,\mathrm{m\,s^{-1}}$
3 $8.32\,\mathrm{N}$, $4.16\,\mathrm{rad\,s^{-1}}$
4 $48\,\mathrm{N}$, $0.476\,\mathrm{m}$
5 (a) $21.6\,\mathrm{N}$ (b) $8.8\,\mathrm{N}$
6 (a) $\dfrac{lg\sin^2\phi}{4\cos\phi}$ **7** $2\sqrt{\left(\dfrac{g}{3l}\right)}$
8 Particle model, $\dfrac{5mg}{3}$, $2\sqrt{(gl)}$
    (b) The tensions in the two parts of the string are the same.
9 $\dfrac{7mg}{12}$ **10** $17:15$ **11** $2\sqrt{\left(\dfrac{g}{3l}\right)}$
12 $9.56\,\mathrm{m\,s^{-1}}$
13 (a) $28.8\,\mathrm{m\,s^{-1}}$ (b) $13.6\,\mathrm{m\,s^{-1}}$
14 (a) 0.373 (b) $13.6\,\mathrm{m\,s^{-1}}$

Exercise 4E

1 (a) $8.40\,\mathrm{m\,s^{-1}}$ (b) $127\,\mathrm{N}$
2 $3.13\,\mathrm{m\,s^{-1}}$ **3** $4.43\,\mathrm{m\,s^{-1}}$, $3.13\,\mathrm{N}$
4 (a) $10.2\,\mathrm{m\,s^{-1}}$ (b) $5.12\,\mathrm{m\,s^{-1}}$ No
5 (b) $2\sqrt{(gl)}$
6 As a particle   (b) $ag(2+\sqrt{3})$
    (c) $0.232\,mg$ towards O
7 $5.64\,\mathrm{m\,s^{-1}}$

Exercise 4F

1 $4.43\,\mathrm{m\,s^{-1}}$, $3.68\,\mathrm{N}$
2 (a) $9.90\,\mathrm{m\,s^{-1}}$ (b) $206\,\mathrm{N}$, $147\,\mathrm{N}$
3 (a) $7.78\,\mathrm{m\,s^{-1}}$ (b) Moves as a projectile until string is taut again.
4 (a) $6.64\,\mathrm{m\,s^{-1}}$ (b) $64.5°$
5 (a) As a particle   (b) $4.18\,\mathrm{m\,s^{-1}}$, $168\,\mathrm{N}$
6 $3mg(1-\cos\theta)$, $109°$
7 $\sqrt{(3gl)}$
8 (a) $\sqrt{(gr)}$, $\frac{1}{2}mg$ (b) $\sqrt{(2gr)}$
9 $\sqrt{(ag)}$
11 (a) Particle   (b) $\sqrt{(ag)}$ (c) $\sqrt{(5ag)}$ (d) $2a$

Exercise 4G

1 $4.47\,\mathrm{rad\,s^{-1}}$
2 (a) $2u$ (b) $\dfrac{(4u^2+2rg)}{3rg}$
3 $18.1°$
4 (a) $65.1°$ (b) $1.44\,\mathrm{m\,s^{-1}}$ (c) $0.298\,\mathrm{m}$
5 (a) $18.4\,\mathrm{N}$ (b) $4.04\,\mathrm{rad\,s^{-1}}$ (c) $0.759\,\mathrm{m}$
6 $4.92\,\mathrm{m\,s^{-1}}$
7 (a) $\sqrt{\left(\dfrac{19gl}{15}\right)}$ (b) $\dfrac{23mg}{30}$
    (c) tension   (d) air resistance, wind
8 (a) $300\,\mathrm{N}$ (b) $78.7°$
    (c) The hammer may be treated as a particle; the wire is regarded as inextensible; air resistance is neglected; arm plus wire taken as a straight line.

Exercise 5A

1 $\left(2,\frac{4}{3}\right)$ **2** $\left(\frac{3}{4},\frac{3}{10}\right)$ **3** $\left(\frac{124}{75},\frac{254}{105}\right)$
4 $\left(\frac{3}{5}b,0\right)$ **5** $\left(\frac{1}{3}a,\frac{1}{3}h\right)$ **6** $\left(\frac{9}{16},\frac{7}{10}\right)$
7 $\left(\frac{4r}{3\pi},\frac{4r}{3\pi}\right)$ **8** $\left(\frac{\pi}{2}-1,\frac{\pi}{8}\right)$ **9** $\left(\frac{1}{e-1},\frac{e+1}{4}\right)$
10 $\left(\frac{28}{3\pi}\right)$ **11** $\left(\frac{1}{2},\frac{2}{5}\right)$

Exercise 5B

1 $(3, 0)$ 2 $(1, 0)$ 3 $(\frac{\pi}{2}, 0)$ 4 $(0, \frac{1}{2})$

5 $(\frac{5}{6}, 0)$ 6 $(\frac{7}{8}, 0)$ 7 $(\frac{5}{8}, 0)$

8 $(2\ln 2, 0)$ 9 $(1\frac{2}{7}, 0)$

10 $\left(\dfrac{1}{2}\dfrac{(e^2+1)}{(e^2-1)}, 0\right)$ 11 $\left(\dfrac{e^2-3}{2(e^2-1)}, 0\right)$

12 $\dfrac{3(h^2+4lh+2l^2)}{4(h+3l)}$ 13 $\sqrt{6}:1$

14 $3\frac{3}{14} = 3.21\,\text{cm}$ 16 $2.33\,\text{cm}$

Exercise 5C

1 $69.4°$ 4 $53.1°$ 5 $30°$

6 $r\sqrt{3}$ 7 $\frac{25}{11}r$ from joint face

10 (b) $\dfrac{mg}{\sqrt{2}}$

Exercise 5D

3 (a) (i) slide $\alpha = 26.6°$, topple $\alpha = 63.4°$

   (ii) slide

   (b) (i) slide $\alpha = 26.6°$, topple $\alpha = 18.4°$

   (ii) topple

4 95.3 5 7.00

6 (b) 588 N (c) slides first

7 (a) $\dfrac{\sqrt{2}\mu\,mg}{1-\mu}$ (b) $\dfrac{mg}{\sqrt{2}}$ (c) $\mu > \frac{1}{3}$

8 (b) $9°$

Review exercise 2

1 $51.2\,\text{N}$, $17.3\,\text{rad s}^{-1}$

3 (a) $mg\sec\beta$ (b) $\dfrac{g}{a\cos\beta}$

4 $0.775\,\text{m}$ 6 $\dfrac{4l}{3}$

8 (i) (a) $m^2g^2 + \dfrac{m^2v^4}{d^2}$ (b) $\dfrac{gd}{v^2}$

   (ii) Particle model, air resistance

9 $19.2\,\text{m s}^{-1}$

10 (a) $\dfrac{3l}{2}$ (c) $2mg$ (d) $2\pi\sqrt{\left(\dfrac{l}{g}\right)}$

12 $1.60\,\text{h}$

13 (a) $\frac{5}{4}mg$ (b) $8ma\omega^2 - \frac{3}{4}mg$

14 (a) $11\,\text{cm}$ (b) $12\,\text{cm}$

15 (a) $3.33\,\text{m s}^{-1}$, $0.566\,\text{m}$

   (b) Ropes may be modelled by a light inextensible string. Weight of seat may be neglected. Air resistance may be neglected.

Acceleration due to gravity is const may be modelled as a particle.

(c) Take into account weight of seat, air resistance, mass of ropes, variation of g, size and shape of child.

16 $T = \dfrac{mV^2}{a} - mg\,(2 - 3\cos\theta)$

$H = \dfrac{27a}{16}$

17 $\dfrac{4a}{9}$, $R = 5\frac{2}{3}mg$

18 (a) $m\sqrt{(5lg)}$ (b) $T = mg$ (c) $\dfrac{l}{3}$

20 $\dfrac{1}{a^2}[u^2 - 2ga(1-\cos\theta)]$

21 $m[\dfrac{u^2}{l} - g(2 - 3\cos\alpha)]$

22 (b) (i) $\sqrt{(3gr)}$ (ii) $\sqrt{(7gr)}$
   (c) (i) $2mg$ (ii) $8mg$

23 $\cos\theta = \frac{2}{3}\cos\beta$

24 (a) $\dfrac{3u}{8}$

26 $\dfrac{5a}{16}$, $\tan\alpha = \frac{16}{45}$

27 $(H+h):(3H-h)$

28 $\theta = 71.5°$

29 (c) $a, \dfrac{2a}{\pi}$ (d) $78°$

32 (b) $30°$

33 (b) $\dfrac{8a}{9}$

34 (a) $\dfrac{2a}{\pi}$ 35 $\dfrac{10a}{3}$ 36 $\dfrac{1}{\sqrt{3}}$

37 (b) $74°$ 38 (b) $50°$

39 (b) (i) topples
   (ii) returns to vertical
   (iii) remains at rest.

40 (b) $38°$ 41 (b) $38°$

Examination style paper M3

1 (a) $1.59\,\text{m}$ (b) $0.0628\,\text{N}$

3 $\mu = 0.43$

4 $x = t^3 + 2t + 3$

5 (a) radius $24\,\text{cm}$ (b) length $25\,\text{cm}$

6 (b) $6.64\,\text{m s}^{-1}$, $2\frac{1}{2}\,\text{m}$ from O

7 (a) $\left(\dfrac{2a}{3}, 0\right)$ (b) $\dfrac{61}{27}a$

List of symbols and notation

The following notation will be used in all Edexcel examinations.

$\in$	is an element of
$\notin$	is not an element of
$\{x_1, x_2, \ldots\}$	the set with elements $x_1, x_2, \ldots$
$\{x : \ldots\}$	the set of all x such that $\ldots$
$n(A)$	the number of elements in set A
$\varnothing$	the empty set
$\mathscr{E}$	the universal set
A'	the complement of the set A
$\mathbb{N}$	the set of natural numbers, $\{1, 2, 3, \ldots\}$
$\mathbb{Z}$	the set of integers, $\{0, \pm 1, \pm 2, \pm 3, \ldots\}$
$\mathbb{Z}^+$	the set of positive integers, $\{1, 2, 3, \ldots\}$
$\mathbb{Z}_n$	the set of integers modulo n, $\{0, 1, 2, \ldots, n - 1\}$
$\mathbb{Q}$	the set of rational numbers $\left\{\dfrac{p}{q} : p \in \mathbb{Z}, q \in \mathbb{Z}^+\right\}$
$\mathbb{Q}^+$	the set of positive rational numbers, $\{x \in \mathbb{Q} : x > 0\}$
$\mathbb{Q}_0^+$	the set of positive rational numbers and zero, $\{x \in \mathbb{Q} : x \geqslant 0\}$
$\mathbb{R}$	the set of real numbers
$\mathbb{R}^+$	the set of positive real numbers, $\{x \in \mathbb{R} : x > 0\}$
$\mathbb{R}_0^+$	the set of positive real numbers and zero, $\{x \in \mathbb{R} : x \geqslant 0\}$
$\mathbb{C}$	the set of complex numbers
(x, y)	the ordered pair x, y
$A \times B$	the cartesian product of sets A and B, $A \times B = \{(a, b) : a \in A, b \in B\}$
$\subseteq$	is a subset of
$\subset$	is a proper subset of
$\cup$	union
$\cap$	intersection
$[a, b]$	the closed interval, $\{x \in \mathbb{R} : a \leqslant x \leqslant b\}$
$[a, b), [a, b[$	the interval $\{x \in \mathbb{R} : a \leqslant x < b\}$
$(a, b], \,]a, b]$	the interval $\{x \in \mathbb{R} : a < x \leqslant b\}$
$(a, b), \,]a, b[$	the open interval $\{x \in \mathbb{R} : a < x < b\}$
$y \, R \, x$	y is related to x by the relation R
$y \sim x$	y is equivalent to x, in the context of some equivalence relation
$=$	is equal to
$\neq$	is not equal to
$\equiv$	is identical to *or* is congruent to

$\approx$	is approximately equal to
$\cong$	is isomorphic to
$\propto$	is proportional to
$<$	is less than
$\leqslant$, $\ngtr$	is less than or equal to, is not greater than
$>$	is greater than
$\geqslant$, $\nless$	is greater than or equal to, is not less than
∞	infinity
$p \wedge q$	p and q
$p \vee q$	p or q (or both)
$\sim p$	not p
$p \Rightarrow q$	p implies q (if p then q)
$p \Leftarrow q$	p is implied by q (if q then p)
$p \Leftrightarrow q$	p implies and is implied by q (p is equivalent to q)
$\exists$	there exists
$\forall$	for all
$a + b$	a plus b
$a - b$	a minus b
$a \times b$, ab, $a.b$	a multiplied by b
$a \div b$, $\dfrac{a}{b}$, a/b	a divided by b
$\displaystyle\sum_{i=1}^{n} a_i$	$a_1 + a_2 + \ldots + a_n$
$\displaystyle\prod_{i=1}^{n} a_i$	$a_1 \times a_2 \times \ldots \times a_n$
$\sqrt{a}$	the positive square root of a
$|a|$	the modulus of a
$n!$	n factorial
$\dbinom{n}{r}$	the binomial coefficient $\dfrac{n!}{r!(n-r)!}$ for $n \in \mathbb{Z}^+$ $\dfrac{n(n-1)\ldots(n-r+1)}{r!}$ for $n \in \mathbb{Q}$
$f(x)$	the value of the function f at x
$f : A \rightarrow B$	f is a function under which each element of set A has an image in set B
$f : x \mapsto y$	the function f maps the element x to the element y
f^{-1}	the inverse function of the function f
$g \circ f$, gf	the composite function of f and g which is defined by $(g \circ f)\,(x)$ or $gf(x) = g(f(x))$
$\displaystyle\lim_{x \to a} f(x)$	the limit of $f(x)$ as x tends to a
Δx, δx	an increment of x
$\dfrac{dy}{dx}$	the derivative of y with respect to x
$\dfrac{d^n y}{dx^n}$	the nth derivative of y with respect to x

$f'(x), f''(x), \ldots, f^{(n)}(x)$	the first, second, $\ldots$, nth derivatives of f(x) with respect to x
$\int y\,dx$	the indefinite integral of y with respect to x
$\int_a^b y\,dx$	the definite integral of y with respect to x between the limits $x = a$ and $x = b$
$\dfrac{\partial V}{\partial x}$	the partial derivative of V with respect to x
$\dot{x}, \ddot{x}, \ldots$	the first, second, $\ldots$ derivatives of x with respect to t
e	base of natural logarithms
e^x, exp x	exponential function of x
$\log_a x$	logarithm to the base a of x
$\ln x$, $\log_e x$	natural logarithm of x
$\lg x$, $\log_{10} x$	logarithm of to base 10
sin, cos, tan cosec, sec, cot	the circular functions
arcsin, arccos, arctan arccosec, arcsec, arccot	the inverse circular functions
sinh, cosh, tanh cosech, sech, coth	the hyperbolic functions
arsinh, arcosh, artanh, arcosech, arsech, arcoth	the inverse hyperbolic functions
i, j	square root of -1
z	a complex number, $z = x + iy$
Re z	the real part of z, Re $z = x$
Im z	the imaginary part of z, Im $z = y$
$|z|$	the modulus of z, $|z| = \sqrt{(x^2+y^2)}$
arg z	the argument of z, arg $z = \arctan\dfrac{y}{x}$
z^*	the complex conjugate of z, $x - iy$
M	a matrix **M**
$\mathbf{M}^{-1}$	the inverse of the matrix **M**
$\mathbf{M}^T$	the transpose of the matrix **M**
det **M**, $|\mathbf{M}|$	the determinant of the square matrix **M**
a	the vector **a**
$\overrightarrow{AB}$	the vector represented in magnitude and direction by the directed line segment AB
$\hat{\mathbf{a}}$	a unit vector in the direction of **a**
i, j, k	unit vectors in the directions of the cartesian coordinate axes
$|\mathbf{a}|$, a	the magnitude of **a**
$|\overrightarrow{AB}|$, AB	the magnitude of $\overrightarrow{AB}$
a . b	the scalar product of **a** and **b**
$\mathbf{a} \times \mathbf{b}$	the vector product of **a** and **b**

A, B, C, etc.	events
$A \cup B$	union of the events A and B
$A \cap B$	intersection of the events A and B
$P(A)$	probability of the event A
A'	complement of the event A
$P(A\|B)$	probability of the event A conditional on the event B
X, Y, R, etc.	random variables
x, y, r, etc.	values of the random variables X, Y, R, etc.
$x_1, x_2, \ldots$	observations
$f_1, f_2, \ldots$	frequencies with which the observations $x_1, x_2, \ldots$ occur
$p(x)$	probability function $P(X = x)$ of the discrete random variable X
$p_1, p_2, \ldots$	probabilities of the values $x_1, x_2, \ldots$ of the discrete random variable X
$f(x), g(x), \ldots$	the value of the probability density function of a continuous random variable X
$F(x), G(x), \ldots$	the value of the (cumulative) distribution function $P(X \leqslant x)$ of a continuous random variable X
$E(X)$	expectation of the random variable X
$E[g(X)]$	expectation of $g(X)$
$Var(X)$	variance of the random variable X
$G(t)$	probability generating function for a random variable which takes the values 0, 1, 2, $\ldots$
$B(n, p)$	binomial distribution with parameters n and p
$N(\mu, \sigma^2)$	normal distribution with mean μ and variance σ^2
μ	population mean
σ^2	population variance
σ	population standard deviation
$\bar{x}, m$	sample mean
$s^2, \hat{\sigma}^2$	unbiased estimate of population variance from a sample,

$$s^2 = \frac{1}{n - 1}\sum(x_i - \bar{x})^2$$

ϕ	probability density function of the standardised normal variable with distribution $N(0, 1)$
Φ	corresponding cumulative distribution function
ρ	product-moment correlation coefficient for a population
r	product-moment correlation coefficient for a sample
$Cov(X, Y)$	covariance of X and Y

Index